THE SMART OFFICE PROJECT

360

SMARTER STACK

DELIVERING INTELLIGENT
WORKPLACES WITH THE '360
SMARTER STACK' METHODOLOGY

NATHAN SHEWRING

The Smart Office Project

First Edition

Contents

Foreword

In an era defined by transformation, uncertainty, and rapid innovation, the workplace has become more than just a place to sit and work. It has become an ecosystem — one that must balance business performance with human experience, operational resilience with environmental sustainability, and cutting-edge technology with practical, scalable delivery.

This book, *The Smart Office Project*, is the natural evolution of my previous work, *The Smart Awakening*, which explored the broader shift toward intelligent living, working, and thinking. Now, we shift gears and focus in on the office environment — the day-to-day operating space for thousands of businesses and millions of people.

The smart office isn't just about putting sensors in ceilings or tablets on meeting rooms. It's about designing spaces and systems that respond to real needs, deliver meaningful value, and are sustainable and scalable for the future. It's about creating a workplace that isn't just connected — but cohesive, context-aware, and intelligently orchestrated.

At the core of this journey is the 360 Smarter Stack — a practical, delivery-oriented framework I've developed to guide businesses, consultants, and technology teams from vision to execution. This book will walk you through each layer of the stack, offering strategic insights, implementation guidance, and real-world considerations.

Whether you're a project sponsor, IT leader, facilities manager, designer, or technologist — I hope this book becomes your go-to reference for smart workplace transformation. Let's build something truly intelligent.

Nathan Shewring
Author, Consultant & Technologist

About the Author

Nathan Shewring is a forward-thinking technology leader, project delivery specialist, and emerging voice in the world of smart buildings and digital transformation. With over a decade of experience delivering IT, AV, and workplace infrastructure projects across complex enterprise environments, Nathan is known for bridging the gap between technical innovation and real-world operational impact.

From leading high-profile refurbishment and downsizing initiatives to pioneering smart access control and AV modernisation programs, Nathan's career has been defined by practical execution, cross-functional collaboration, and a deep understanding of how people interact with spaces.

Blending his background in finance, IT project management, and a growing interest in IoT and smart systems, Nathan brings a rare 360-degree perspective to the built environment. His passion lies in making spaces smarter — not just through technology, but through purpose-led design, data-driven decisions, and human-centric engagement.

He is also the founder of NKS Digital Media, a platform exploring immersive tech, high-end display systems, and digital experiences that "wow" — both in corporate settings and beyond.

This book reflects Nathan's belief that smart buildings aren't just about sensors and software — they're about vision, value, and making better environments for people to thrive. Through clear frameworks like the Smarter Stack and real-world use cases, he offers a roadmap for teams ready to lead the next era of intelligent workplaces.

When not driving transformation in the workplace tech space, Nathan enjoys coaching youth football, studying accounting to expand his strategic toolkit, and exploring new ways to bring technology and creativity together.

The Smart Office Project

Introduction: Why smarter offices matter now

The modern office is under pressure like never before. Hybrid work, climate goals, operational efficiency, employee wellbeing, and digital transformation are all converging. Legacy office environments — once simply functional — are now being reimagined as dynamic, data-driven, people-centric spaces.

Here's why smart offices matter more than ever:

1. The New Hybrid Workforce

The workplace is no longer a static 9–5 destination. It's now a flexible, on-demand service that must adapt to the ebb and flow of hybrid schedules, hot desking, virtual meetings, and shared collaboration zones. Smart offices enable seamless integration of digital tools, booking systems, and real-time occupancy data, making hybrid work intuitive, not frustrating.

2. Environmental, Social, and Governance (ESG) Pressures

Sustainability isn't optional — it's business-critical. Buildings account for roughly 40% of global energy use. Smart offices give organizations the tools to measure, reduce, and optimize energy usage, water consumption, and carbon footprint — while supporting wellbeing and inclusivity initiatives through intelligent design.

3. Operational Efficiency & Cost Control

Energy costs, underused space, and inefficient systems eat into the bottom line. A smart building can automate, monitor, and optimize everything from HVAC to lighting to cleaning schedules. The result? Lower operational costs, fewer manual interventions, and more responsive environments.

4. Employee Experience as a Differentiator

Attracting and retaining talent now depends heavily on workspace quality. People expect intuitive, tech-enabled, comfortable environments. The smart office delivers on this by aligning technology and space around user needs — from air quality to acoustics to digital wayfinding.

5. Security, Safety & Compliance

Post-pandemic offices must ensure safe access, real-time occupancy tracking, and flexible alerting systems — while also adhering to cybersecurity, GDPR, and building safety regulations. Smart systems offer the automation and transparency needed to meet those obligations with confidence.

6. Data-Driven Decision Making

A truly smart office generates insight. Every sensor, booking, access point, or system event becomes a datapoint that — when correctly integrated — drives better decisions. From redesigning floor plans to justifying capital spend, smart data creates visibility that was never previously possible.

- Evolution from traditional fit-outs to tech-enabled, data-driven workplaces

For decades, office fit-outs followed a familiar script: desks, partitions, cabling, lighting, and HVAC — all designed to meet static needs, based on a one-size-fits-all layout. Technology, where present, was layered on top rather than built in. Workplace systems were siloed. Once complete, a fit-out was considered "done," with little thought given to how the space might evolve or respond to change.

But the nature of work — and our relationship with the office — has changed dramatically.

Traditional Fit-Outs: Function Over Flexibility

In a conventional fit-out, decisions were driven primarily by physical space planning and compliance:

- Capacity planning based on headcount ratios
- Generic meeting rooms with basic AV
- Limited real-time visibility into usage or performance
- Manual processes for room booking, maintenance, and access control
- Isolated systems (lighting, HVAC, security) with little or no integration

Once built, these environments remained largely static — with reconfigurations often costly and disruptive.

The Shift: From Static Shells to Responsive Systems

Today, the office is expected to adapt to users, workflows, and broader business dynamics. This shift is underpinned by three key changes:

Technology is no longer peripheral — it's embedded.

Systems are now planned into the architectural blueprint: smart lighting, sensor networks, integrated AV, building management systems, and workplace apps are all considered early, not late.

Spaces are expected to sense, respond, and optimise.

With the rise of IoT and connected systems, modern offices collect real-time data on occupancy, air quality, noise, energy usage, and user behaviour — enabling dynamic responses and predictive adjustments.

Office Fit-Outs are now part of a longer digital lifecycle.

A smart fit-out is not the end — it's the beginning of a data-driven relationship with the space. Configuration, usage, and performance can all evolve based on analytics and user feedback.

From Fit-Out to Fit-For-Future

This evolution has reframed what "fit-for-purpose" really means. It's no longer just about desks per square metre or air changes per hour — it's about delivering adaptable, efficient, and people-centric environments. **Fit-outs must now accommodate:**

- Hybrid work and fluctuating occupancy
- Real-time visibility into asset performance
- Seamless user experience across physical and digital interfaces
- Integration with wider enterprise systems (HR, IT, ESG platforms)
- Long-term flexibility without starting from scratch

Why This Evolution Matters

As lease terms shorten, environmental standards tighten, and employee expectations rise, the traditional approach to fit-outs simply isn't enough. Organizations need offices that are as intelligent and flexible as the people who use them.

Smart offices are the natural evolution — not an upgrade or add-on, but a new way of delivering value from the workplace itself.

- Introduction to the **360 Smarter Stack** as a structured delivery model

As the office has evolved from a static fit-out to a responsive digital platform, so too must the way we design, deliver, and operate these spaces. Traditional project methodologies — focused largely on construction, IT rollout, or isolated systems integration — often fall short when applied to smart environments. That's where the 360 Smarter Stack comes in.

A New Approach for a New Workplace

The **360 Smarter Stack** was developed as a comprehensive delivery framework to bring structure and clarity to the complex process of creating smart office environments. Inspired by early models such as those discussed by the Monday Live community, this framework has been refined and shaped through practical application in real-world projects.

The name "360" reflects its core principle: providing a complete, end-to-end view of all the critical domains that must be considered for a successful smart office transformation — from strategic intent at the top, down to the physical infrastructure that underpins it all. The Stack breaks the smart workplace into eight interconnected layers, each one addressing a vital aspect of strategy, design, technology, integration, and operations.

It's more than a conceptual model — the **360 Smarter Stack** is a practical, layered methodology designed to ensure every project is grounded in purpose, driven by data, and delivered with cross-disciplinary alignment.

The Stack breaks down the smart office into eight interdependent layers, each representing a critical domain — from strategic vision to physical infrastructure. It's designed to bring clarity, structure, and consistency to what is often a fragmented and complex delivery process.

Where many projects fail due to misalignment between teams, poorly integrated systems, or missing business value, the Smarter Stack ensures every stakeholder — from IT to Facilities, from exec sponsor to end user — can work to a common model with clear touchpoints and outcomes.

The 360 Smarter Stack should be viewed as a practical delivery methodology, a benchmark for smart transformation, and the backbone of this book.

The Eight Layers of the 360 Smarter Stack

Layer	Core Focus	Why It Matters
1. Vision & Value	ESG, ROI, Experience	Aligns projects with strategic business outcomes
2. Operations & Culture	FM, IT, SLAs, training	Ensures solutions are adoptable and sustainable

3. User Experience	Apps, signage, physical touchpoints	Drives employee engagement and workplace satisfaction
4. Applications & Services	Booking, analytics, security features	Powers the visible, functional smart tools
5. Data & Context	Semantic modelling, trends, history	Turns raw data into insight and action
6. Integration Layer	APIs, middleware, orchestration	Enables cross-system automation and workflow
7. Systems & Devices	HVAC, lighting, AV, sensors	Provides environmental control and telemetry
8. Physical Environment	Cabling, racks, electrical capacity	The physical layer everything else depends on

Why It Works

The 360 Smarter Stack is designed to address both the "what" and the "how". It maps not only the components of a smart workplace but also the sequence, dependencies, and governance models required to bring it all together.

Each layer builds upon the one beneath it:

- You can't extract value from data (Layer 5) without integrating systems (Layer 6).

- You can't deliver seamless apps and UX (Layers 3–4) unless the underlying infrastructure (Layers 7–8) is solid.

- You can't build sustainable, long-term success unless you've aligned the entire project with strategic vision and operational maturity (Layers 1–2).
- This layered approach means that risks are managed early, silos are broken down, and the project evolves logically from concept to commissioning to continuous improvement.

The Benchmark for Smart Office Delivery

The 360 Smarter Stack has already informed successful smart office projects by offering a shared language and roadmap that cuts across design, IT, facilities, and executive teams. It helps to:

- Bridge strategic goals and technical delivery
- Avoid redundant or misaligned investments
- Ensure interoperability from day one
- Enable clear scoping, phasing, and prioritisation

As we move through the book, each chapter will explore one layer of the Stack in detail — offering frameworks, checklists, real-world insights, and lessons learned to help you turn your smart office ambitions into a fully functioning reality.

Part 1: The Strategic Imperative

Chapter 1: Introducing the 360 Smarter Stack

In the journey to deliver a truly smart workplace, there's a common trap: jumping straight into technology decisions without a clear structure to guide the process. This often results in fragmented systems, misaligned goals, or solutions that are difficult to support long-term. To avoid this, we need a model that not only maps what needs to be delivered, but also shows how all the elements work together. The 360 Smarter Stack does exactly that — providing a clear, layered framework that connects vision to infrastructure, people to systems, and data to value.

Overview of the 8 layers

The 360 Smarter Stack is designed as a layered delivery model that brings structure, logic, and cohesion to smart office transformation. Each layer represents a domain of responsibility, a set of outcomes, and a key consideration in both planning and delivery. Critically, the stack is not linear — while there's a natural order of dependency, most successful projects involve collaboration and iteration across layers throughout the lifecycle.

Think of the Stack as a full-circle view (360 degrees) — ensuring no crucial element is overlooked in strategy, technology, operations, or experience.

◆ **Layer 1: Vision & Value**

Core Focus: Strategic intent, ROI, ESG alignment, user needs

This top layer defines why the project exists. It's about articulating a clear vision that aligns with business priorities — whether that's decarbonisation, hybrid working, operational efficiency, or employee wellbeing. It includes identifying key drivers, setting measurable objectives, and gaining stakeholder buy-in. Without clarity at this level, even technically perfect solutions can underdeliver.

◆ Layer 2: Operations & Culture

Core Focus: Facilities, IT, support models, service readiness

Smart technology needs smart operations behind it. This layer focuses on the people and processes needed to operate, support, and evolve the smart workplace. It looks at service models, FM and IT collaboration, skills gaps, training needs, and cultural readiness. Projects often stumble when operational impact is considered too late — this layer ensures it's embedded from the outset.

◆ Layer 3: User Experience

Core Focus: Digital and physical touchpoints, journey design, accessibility

This is the layer employees and visitors interact with daily — signage, booking systems, mobile apps, room panels, kiosks, or even how lighting and temperature feel. A well-designed experience drives engagement and satisfaction, while a poor one undermines adoption. This layer ensures the workplace is not just functional, but frictionless, intuitive, and human-centred.

◆ Layer 4: Applications & Services

Core Focus: Smart office tools, integrations, automation workflows

These are the applications that deliver the functionality of a smart environment — space booking, environmental dashboards, digital twins, AV control, security apps, occupancy tools. This layer involves selecting, configuring, and integrating these tools so they work as a cohesive suite, rather than isolated islands of functionality.

◆ Layer 5: Data & Context

Core Focus: Semantics, data modelling, real-time vs historical insights

This layer is where information becomes intelligence. It involves structuring and modelling the data generated across systems (e.g. sensors, apps, BMS) to create meaningful insights. When done well, it unlocks predictive maintenance, usage trend analysis, and continuous optimisation. When done poorly, data becomes overwhelming or underused.

◆ Layer 6: Integration Layer

Core Focus: APIs, middleware, data pipelines, orchestration

Sitting beneath the data and services layers, this is the technical glue that ensures systems can talk to each other. It's what allows a meeting room booking to trigger lighting, HVAC, and AV setup — or enables unified data visualisation across different platforms. Integration is often underestimated but is essential for delivering a cohesive, automated experience.

◆ Layer 7: Systems & Devices

Core Focus: Smart hardware — sensors, AV, HVAC interfaces, access control

This is where the digital meets the physical. It includes all the deployed systems and IoT devices that collect data, control environments, and enable automation. From lighting and blinds to people counters and air quality monitors, this layer is about specifying and deploying technology that serves the upper layers of the stack.

◆ Layer 8: Physical Environment

Core Focus: Power, cabling, connectivity, racks, PoE infrastructure

The foundation of any smart office is the infrastructure it runs on. Often overlooked, this layer covers all physical enablers — structured cabling, power supply, containment, network access, and space for hardware (racks, patch panels, sensor placement). It must be aligned with the rest of the stack early to prevent costly rework later.

How the Layers Interrelate in Real-World Delivery

In theory, layers can be discussed individually. In practice, they never exist in isolation.

One of the defining strengths of the 360 Smarter Stack is that it recognises the interdependencies between strategic goals, operational processes, user expectations, and technical systems — and turns those interdependencies into a strength rather than a weakness. Successful smart office delivery is not a sequence of siloed activities, but rather a coordinated choreography across all eight layers.

Below is a deeper look at how the layers interact in real-world projects — and why recognising these relationships early can make or break a smart workplace transformation.

From Vision to Infrastructure: Strategy Must Shape Delivery

Smart office projects often begin with a compelling idea — improve the hybrid work experience, meet ESG targets, reduce operational costs, or modernise ageing infrastructure. But when those ideas are handed off without structure, the delivery team may focus too soon on devices or software, disconnected from the original intent.

The 360 Smarter Stack ensures that Layer 1: Vision & Value drives decisions all the way down to Layer 8: Physical Environment. For example:

A goal of carbon reduction informs sensor choices (Layer 7), energy dashboards (Layer 4), and data modelling (Layer 5).

A vision for inclusive, accessible workspaces shapes UX design (Layer 3), signage and apps (Layer 4), and even equipment placement (Layer 8).

Each strategic aim must be translated into technical requirements, supporting processes, and physical constraints across the stack. If this isn't done deliberately, vision leaks through the cracks.

UX Doesn't Sit on Top — It Lives Throughout the Stack

In many projects, user experience is treated as a surface-level layer — something to make "look good" at the end. But in reality, Layer 3: User Experience is where multiple layers converge.

Take a smart meeting room:

The booking experience (Layer 4) relies on data from occupancy sensors (Layer 7), pulled through an API (Layer 6), and visualised in a mobile app or panel (Layer 3).

The comfort of the space is controlled via HVAC and lighting systems (Layer 7), whose settings may change based on user profiles or preferences stored in applications (Layer 4) and analysed through data models (Layer 5).

Every click, swipe, and interaction is the tip of an iceberg, with functionality dependent on well-orchestrated integration across the stack. UX success or failure is almost always downstream of decisions made across Layers 4–8.

Integration Is the Keystone Layer

Layer 6: Integration is often the quiet hero — or the silent saboteur — of smart office projects.

In real-world delivery:

Systems must pass data seamlessly — from access control to lighting, from sensors to analytics platforms.

Workflows such as "employee books a desk, receives a QR code, HVAC adjusts, welcome message appears" require real-time orchestration and event-driven logic.

Without a robust integration layer:

- Applications become siloed and frustrating.
- Data is inconsistent or duplicated.
- Automation is limited or brittle.

This layer enables everything above it to feel connected and intelligent, and everything below it to be leveraged for real value. It's the glue that binds the stack into a functional whole.

Data & Context: More Than Collection — It's Translation

Layer 5: Data & Context bridges the raw data coming from devices (Layer 7) and the meaningful insight needed at the business level (Layer 1).

In practice:

Facilities teams may want to know which zones are underused — requiring sensor data (Layer 7), pulled via integration (Layer 6), and modelled into usable formats (Layer 5).

Leadership may want to compare CO_2 levels across offices globally — demanding consistent data standards, semantic tagging, and normalised historical views.

This layer is the translator — it converts infrastructure outputs into language the business can act on. It also ensures the same data can be used for both operational control and strategic planning.

Operations Must Be Baked In — Not Bolted On

A common failure point in smart building delivery is ignoring Layer 2: Operations & Culture until the end of the project. But if the operations team can't support or maintain the new technology, the system quickly degrades.

In real-world delivery:

If FM teams aren't trained on the sensor network, alerts may be ignored or misunderstood.

If IT isn't prepared to support the booking platform, users lose trust and revert to manual processes.

If service models don't align with new tech (e.g. managing data pipelines or APIs), the integrations break down silently.

This layer ensures that what is built can be operated, governed, and improved over time. It should be engaged from the start, not treated as a go-live afterthought.

Physical Environment Enables — or Constrains — All Above

Layer 8: Physical Environment may sit at the bottom, but it determines what's possible across the entire stack.

For example:

Lack of structured cabling or PoE limits where sensors or panels can be deployed (impacting UX and data collection).

Overloaded switchboards or lack of cooling capacity restrict AV or IT upgrades.

Poor equipment room planning leads to congested patch panels or compromised security.

Smart functionality can only be layered onto an environment that's been designed to support it. Coordinating early between architects, IT, and system designers prevents costly clashes and retrofits.

Smart Delivery Is Not Linear — It's Layer-Aware

While there's a logical flow (from strategy to infrastructure), smart delivery is rarely sequential. In reality:

- Vision and requirements evolve as technology becomes clearer.

- Early infrastructure decisions (like network topology) impact application options later.

- Data modelling informs UX decisions before devices are fully installed.

That's why the 360 Smarter Stack encourages cross-functional dialogue, iterative delivery, and layer awareness — so every team knows how their work impacts the broader outcome.

Summary: Interconnection Is the Real Intelligence

The power of a smart office doesn't lie in any one layer — it emerges from how the layers work together. True intelligence comes not from the devices or dashboards alone, but from a well-integrated stack where:

- Vision guides every decision

- Operations are equipped to support what's built

- Systems and data serve meaningful, human outcomes

- Physical and digital environments are planned as one

In the chapters that follow, we'll unpack each layer in depth — not in isolation, but always with awareness of how it fits within the whole.

Benefits of a Layered Approach to Smart Transformation

Transforming a traditional workplace into a smart, connected environment can quickly become overwhelming. With so many interwoven domains — from cabling to culture, APIs to analytics — it's easy to lose sight of the bigger picture or to focus too heavily on one area at the expense of another.

The 360 Smarter Stack introduces a layered methodology that brings structure, clarity, and control to the chaos. By organising transformation across eight interdependent layers, it helps project teams and stakeholders to navigate complexity, avoid misalignment, and deliver lasting value.

Here are the key benefits of adopting a layered approach:

Clarity Across Stakeholder Groups

In most smart workplace projects, stakeholders come from diverse domains: IT, FM, HR, Real Estate, Sustainability, Change Management, and more. Each brings a different perspective and set of priorities.

The 360 Smarter Stack gives all parties a shared language and framework:

- Business leaders can focus on Vision & Value and User Experience

- FM and IT teams engage with Operations, Systems, and Integration

- Technology and design partners align with Applications, Data, and Infrastructure

This removes ambiguity, fosters collaboration, and ensures that everyone understands how their part fits into the whole — a vital step in breaking down silos that often derail these projects.

Better Planning and Sequencing

Smart workplace projects involve multiple disciplines working in parallel. Without a layered model, the sequence of design, procurement, installation, and testing can become muddled — leading to rework, delays, or incompatible systems.

A layered approach provides a structured planning tool, allowing teams to:

- Identify dependencies between layers (e.g., APIs needed before apps are built)

- Sequence work logically (e.g., ensure infrastructure is in place before device installation)

- Avoid "tech last" or "tech too early" pitfalls

This improves coordination, de-risks timelines, and ensures smoother handoffs between disciplines.

Strategic Alignment from Start to Finish

One of the biggest risks in smart projects is strategic drift — where initial goals around ESG, user experience, or innovation get diluted or lost as the project progresses.

With a layered model:

- The top layer (Vision & Value) remains anchored as the north star

- Every decision at lower layers is traceable back to strategic intent

- Delivery teams stay focused not just on "what" they're building, but "why"

This keeps outcomes meaningful, aligned to the business case, and easier to measure post-implementation.

Built-In Flexibility and Scalability

Workplaces are evolving — and so are the technologies that support them. The layered approach supports modular growth and future proofing by clearly separating concerns.

For example:

- You can replace an app (Layer 4) without rebuilding the sensor network (Layer 7)

- New data sources can be added to the model (Layer 5) without disrupting existing operations

- Future enhancements like AI or workplace personalisation can plug into the stack with minimal rework

- This decoupling of layers enables evolution without disruption, which is crucial for long-term success.

Improved Risk Management

Complex smart projects are full of risk: technical failure, stakeholder misalignment, poor user adoption, or unexpected integration issues.

The 360 Smarter Stack acts as a risk detection and mitigation tool:

- Gaps in planning become visible earlier (e.g., overlooked training needs or un-scoped integration)

- Impacts can be assessed layer by layer (e.g., how a change in power routing affects systems above)

- Contingencies and dependencies can be mapped with more confidence

- This reduces surprises during delivery and improves project governance.

Holistic Measurement of Success

Traditionally, success is measured by individual KPIs — energy usage, helpdesk tickets, or user surveys. But a layered approach enables multi-dimensional measurement:

- Did the environment achieve the strategic goals (Layer 1)?

- Is the solution operationally sustainable (Layer 2)?

- Are users engaging effectively (Layer 3)?

- Is the data flowing and being used intelligently (Layers 5–6)?

By evaluating outcomes across all eight layers, you build a more complete picture of project performance and can more easily identify where improvements are needed post-deployment.

Accelerated Learning and Reusability

Once a layered approach is used in one office or location, it becomes a template that can be replicated and refined for future sites or use cases.

This unlocks:

- Faster planning cycles on subsequent projects

- Consistency across the portfolio

- A growing knowledge base for internal teams and partners

Ultimately, the 360 Smarter Stack becomes a strategic asset, not just a project tool.

Summary: Turning Complexity Into Confidence

The journey to a smart, connected workplace involves more than just technology. It requires joined-up thinking across business, operations, user needs, and systems — and a structured way to navigate that journey.

By using a layered model:

- You manage complexity with greater confidence

- You align every team to the same strategic outcomes

- You create a flexible, scalable, and resilient foundation for the future

In the next chapters, we'll explore each of the eight layers in detail — starting at the top, where all transformation begins: Vision & Value.

Part 2: Layer-by-Layer Delivery

Layer 1: Vision & Value

Chapter 2: Defining the Smart Office Vision

A smart office isn't built from technology alone — it starts with purpose. Before any sensors are specified, any data model scoped, or any apps designed, the organisation must define what it wants to achieve from its transformation.

Too often, smart workplace initiatives begin with a shopping list of features or a wish to "modernise" — but without a clearly articulated vision, these efforts risk becoming fragmented, underused, or misaligned with wider goals.

This chapter focuses on how to build that vision: rooted in ESG priorities, employee experience, and organisational performance — while also gaining the leadership support and business case clarity needed to carry it through.

We begin with a critical foundation — aligning the smart office with your Environmental, Social, and Governance (ESG) goals, along with wellbeing and productivity drivers.

Aligning with ESG, Wellbeing, and Productivity Outcomes

The modern workplace is no longer just a location — it's a strategic asset in delivering on a company's environmental, social, and operational commitments. A smart office must not only function well, but it must also help people thrive while supporting broader planet-positive objectives.

Let's explore how these pillars come together:

Environmental Goals (E of ESG)

Smart workplaces can play a leading role in helping organisations reduce their environmental impact — and critically, track progress against those ambitions.

Smart strategies should support:

- Energy efficiency through intelligent HVAC, lighting, and occupancy-driven systems

- Carbon footprint reduction via space utilisation data, hybrid working enablement, and remote collaboration tech

- Waste reduction through digital signage, smart printing, and real-time monitoring of consumables

- Sustainable sourcing of materials and devices

More importantly, smart systems help measure and report ESG progress — feeding dashboards, ESG reports, and regulatory returns with real data, not assumptions.

Social Responsibility & Wellbeing (S of ESG)

Employee wellbeing is no longer a 'nice to have' — it's a core driver of engagement, productivity, and retention. The smart office can directly support this in measurable ways.

Examples include:

- Air quality and thermal comfort monitoring and management

- Noise and light level optimisation using responsive systems

- Wellbeing-oriented design cues like digital biophilia, calming environments, and choice of work settings

- Inclusive technologies such as voice control, adaptive lighting, or wayfinding for neurodiverse users

A smart workplace should anticipate and adapt to human needs, not just technical ones. Embedding wellbeing outcomes into your smart vision shows a commitment to your people and helps ensure the investment delivers tangible value beyond efficiency.

Governance & Productivity Outcomes (G of ESG + Business Value)

Governance, in the smart office context, means ensuring systems are:

- Secure

- Ethically deployed (e.g., people-aware data collection)

- Compliant with internal and external standards (ISO, GDPR, etc.)

Smart workplaces also offer new levels of operational intelligence, giving leaders the ability to make evidence-based decisions about:

- Real estate strategy (downsizing, redesign, consolidation)

- Meeting room usage and collaboration patterns

- Headcount distribution vs. workspace needs

- Service delivery (cleaning, catering, IT support) based on real usage

All these support productivity but also enable a more agile and responsive workplace model.

The Big Picture: A Purpose-Driven Smart Office

By rooting your smart office vision in ESG, wellbeing, and productivity, you move from "tech for tech's sake" to a purpose-led transformation.

When well aligned:

- Smart systems support your net zero goals

- Technology actively enhances the human experience

- Data becomes a driver for better governance and smarter operations

This clarity is essential for stakeholder buy-in, funding approval, and long-term support. It also ensures that as you progress through the layers of the 360 Smarter Stack, your decisions stay anchored to what truly matters.

Establishing clear ROI expectations and KPIs

While the term "smart office" conjures images of cutting-edge technology and futuristic experiences, senior decision-makers ultimately want to know one thing: What is the return on investment?

To build trust in the smart transformation and secure long-term support, it's essential to translate ambition into tangible, measurable outcomes. This means establishing clear ROI models and a framework of key performance indicators (KPIs) from the very beginning.

Let's explore how this can be done effectively.

The Challenge of Smart Office ROI

Traditional office fit-outs typically focus ROI on capital costs, lease efficiency, or headcount accommodation. But a smart workplace goes beyond square footage — it touches:

- Experience and engagement

- Efficiency and automation

- Space utilisation

- Carbon footprint

- Operational resilience

This makes ROI multi-dimensional, often requiring both quantitative and qualitative measures — including both direct financial gains and indirect business value.

A strong ROI model answers:

- What outcomes will this transformation unlock?

- How will we know it's working?

- Over what timeframe will the benefits emerge?

Key ROI Categories to Consider

Here are some common categories smart office projects can be measured against:

- Space Optimisation

- Reduction in underutilised space

- Desk-to-headcount ratio improvement

- Energy use per square metre

- Operational Efficiency

- Reduction in manual interventions (e.g. FM callouts)

- Faster issue resolution due to smart monitoring

- Consolidation of redundant systems or vendors

- Employee Experience & Productivity

- Time saved through intuitive booking or wayfinding

- Increased collaboration room utilisation

- Measurable improvements in employee engagement or satisfaction surveys

- Sustainability & ESG Impact

- Energy savings from automated systems

- Data to support net zero reporting

- Reduced carbon impact through flexible working or better space planning

- Technology ROI

- Reduced cost of ownership via centralised management

- Longer lifecycle for tech through modular design

- Vendor consolidation and license cost reduction

Not all of these will apply to every project — but defining which are most relevant helps frame expectations and track results.

Defining KPIs That Matter

A smart office without KPIs is like a business strategy without metrics — it can't be governed, improved, or defended.

Recommended smart office KPIs might include:

- Average daily occupancy rate vs. capacity

- Percentage of meeting room availability

- Energy consumed per employee

- Time taken to resolve FM issues
- User satisfaction with room booking or hybrid tools

- Air quality thresholds met over time

- Uptime of smart systems (AV, sensors, controls)

KPIs should be:

- Aligned to business goals

- Tracked over time

- Reported to stakeholders in accessible formats

- Used to refine both the space and the systems post-deployment

From Investment to Intelligence

The smart workplace is not a one-off cost — it is an investment in continuous intelligence. By linking outcomes to ROI and building in live KPIs, you gain:

- Greater business case clarity

- More targeted technology decisions

- Stronger stakeholder confidence

- A performance-driven roadmap for future improvements

In the next section, we'll explore the importance of bringing senior stakeholders on board early — not just for approval, but to ensure shared ownership of the smart workplace vision and its long-term success.

Gaining Senior Stakeholder Alignment

A smart office project isn't just an IT or Facilities initiative. It's a strategic investment that touches multiple areas of the business — from HR and Operations to Risk, Finance, and ESG. For the project to succeed, it's critical to ensure senior stakeholders across functions are not only informed but aligned, engaged, and supportive from the outset.

Without that alignment, projects risk stalling, becoming siloed, or failing to deliver on their intended outcomes.

Why Senior Alignment Matters

Senior stakeholder buy-in isn't just about budget approval — it creates the conditions for:

- Cross-functional collaboration

- Strategic clarity and prioritisation

- Faster decision-making

- Better resource allocation

- Change management support

Smart offices often require operational changes — new ways of booking desks, new access systems, more data-led facilities decisions. These can't be effectively rolled out if leadership isn't aligned and visibly supportive.

Mapping the Stakeholder Landscape

To align stakeholders, you must first understand who they are and what matters to them. Every smart office programme should begin by mapping internal decision-makers and influencers, such as:

Role	What They Care About
CFO / Finance	ROI, cost efficiency, capital planning
COO / Operations	Business continuity, service performance
HR / People Teams	Employee wellbeing, engagement, hybrid support
CIO / CTO	System integration, data governance, cybersecurity
Heads of Property / Workplace	Utilisation, flexibility, space strategy
ESG / Sustainability Leads	Carbon impact, reporting, compliance

Tailoring the messaging for each group is essential. For example, the ESG team may value the building's smart meters and energy dashboard, while HR will be more interested in how the space enhances wellbeing and collaboration.

How to Build Consensus Early

- Establish a Strategic Steering Group

Bring together key stakeholders to form a smart office steering group. This allows shared ownership and ongoing engagement throughout the programme.

- Co-create the Vision

Invite input on what success looks like. Use workshops, discovery sessions, and structured interviews to understand each stakeholder's priorities.

- Use the 360 Smarter Stack to Communicate

The 360 Smarter Stack provides a common language. It helps stakeholders see how their interests align across layers — from high-level vision and user experience down to infrastructure and data.

- Align on KPIs and Business Case

Set clear success criteria that reflect the interests of all parties. This helps avoid misaligned expectations down the line.

- Ensure Visibility and Regular Touchpoints

Keep stakeholders engaged through regular progress updates, design reviews, and visible wins. Celebrate milestones and communicate what's changing and why.

Unlocking Long-Term Support

Securing alignment upfront is only the first step. Maintaining it throughout the lifecycle of the project — and beyond go-live — ensures the smart office becomes a strategic asset, not a one-off project.

When senior leaders are invested in the outcome:

- Adoption increases

- Cultural shifts are embraced

- Budgets are protected

- Future innovation becomes easier

The most successful smart office transformations are not those with the most tech —
they're the ones where leaders align around a shared vision and stay committed to
realising it.

Chapter 3: Building the Business Case

No matter how compelling the vision or transformative the technology, a smart office initiative must stand up to scrutiny in the boardroom. Business cases are the gateway to funding, executive buy-in, and prioritisation against other competing projects. But traditional ROI models often fall short when applied to smart buildings — they don't capture the full spectrum of benefits, nor do they account for intangible gains like employee satisfaction, ESG impact, or digital readiness.

In this chapter, we explore how to build a robust and multi-dimensional business case, backed by both numbers and narrative, that speaks the language of decision-makers while remaining grounded in long-term strategic value.

Cost/Benefit Modelling

Cost/benefit modelling is the financial core of the business case — a structured, data-backed analysis that compares the required investment against the measurable and perceived value it delivers. For smart office projects, this isn't a straightforward exercise. Unlike traditional infrastructure upgrades, smart environments offer blended returns: operational cost savings, people-centric benefits, ESG outcomes, and long-term strategic positioning. Capturing this full value requires moving beyond linear spreadsheets and embracing a multi-dimensional model.

1. Direct Financial Benefits (Hard ROI)

These are the most traditionally accepted benefits — the ones that appear directly on the P&L or balance sheet.

Energy Efficiency Savings:

Intelligent lighting, HVAC zoning, and occupancy-based control systems can lead to 10–30% reductions in energy consumption. Energy cost savings are quantifiable, often immediate, and contribute to ESG goals.

Real Estate Optimisation:

Smart space analytics can highlight underutilised areas, allowing organisations to consolidate space or avoid costly expansions, leading to significant lease and operational savings.

Reduced Operational Costs:

Predictive maintenance and automated monitoring reduce reactive callouts, extend asset lifespans, and lower FM contract costs. Smart sensors can even inform when bins need emptying or rooms need cleaning — reducing unnecessary labour.

Improved Asset Lifecycle Management:

With real-time usage and health data, major assets like HVAC systems or AV equipment can be better managed, resulting in lower replacement costs and downtime.

2. Indirect Benefits (Soft ROI)

These aren't always easy to quantify — but they are often where real, long-term value lives.

Increased Productivity:

Smart workplaces remove friction — from booking meeting rooms to finding colleagues — saving minutes every day per employee. At scale, that time compounds into meaningful productivity gains.

Improved Employee Experience:

Seamless workspaces, air quality, lighting comfort, and user-friendly apps lead to higher satisfaction, reduced churn, and stronger employer branding — all of which reduce recruitment and onboarding costs.

Hybrid Work Enablement:

Smart offices enable seamless desk/room booking, occupancy visibility, and collaboration tech, making hybrid working work — which in turn allows for longer-term workplace flexibility and cost management.

Brand and Client Perception:

A smart office can be a differentiator in client tours, recruitment, and media — reflecting innovation and investment in people and technology.

3. ESG and Compliance Outcomes

A properly modelled business case should also highlight:

- Carbon reduction metrics, with estimates for tonnes of CO_2 saved annually

- Alignment with internal ESG policies, investor criteria, or public commitments

- Data availability for compliance reporting under frameworks like SECR or TCFD

- Improved indoor air quality (IAQ) for WELL or RESET certification pathways

These translate into reputational, operational, and legal value — often reducing future risks or unlocking ESG-related funding.

4. Structuring the Model

To build a credible and comprehensive model:

Use Baselines:

Establish current-state costs — e.g., energy, lease, FM, churn, productivity benchmarks.

Map Investment Categories:

Break down costs into:

- CapEx: design fees, smart devices, infrastructure upgrades, software licenses

- OpEx: support, data storage, subscription models, analytics services

Model Time-Based Scenarios:

Project returns over 1, 3, and 5 years, showing cumulative savings vs. investment.

Include Intangible Narratives:

While not quantifiable, benefits like "employee experience uplift" should still be named and explained, especially if aligned to HR, wellbeing, or ESG goals.

Prepare Sensitivity Models:

Include scenarios where adoption is slower, costs are higher, or results take longer. This builds trust and realism in the business case.

Example Summary Table

Benefit Area	Description	Estimated Annual Saving	Confidence Level
Energy Efficiency	20% reduction in electricity and gas	£120,000	High
FM Cost Reductions	Smart cleaning and maintenance scheduling	£75,000	Medium
Space Efficiency	Reduce footprint by one floor	£300,000	High
Productivity Gain	2% increase in workforce productivity	£500,000 (soft)	Medium
ESG/Carbon Offsets	200T CO_2 annual reduction	Strategic value	Medium

By grounding your case in numbers, stories, and outcomes, you'll move from "tech proposal" to strategic investment pitch — the kind that resonates at the C-level.

Risk Mitigation

Smart office projects introduce powerful capabilities, but they also come with complexity, integration challenges, and stakeholder expectations. A well-constructed business case must acknowledge these risks — not to deter progress, but to demonstrate preparedness and resilience.

Risk mitigation isn't about eliminating risk entirely. It's about identifying potential pitfalls early, putting controls in place, and reassuring stakeholders that issues won't derail success. When done properly, it becomes a point of strength and confidence, not fear.

1. Technical Risk Mitigation

Smart office systems involve many moving parts — IoT devices, APIs, software platforms, and legacy infrastructure. Without a cohesive plan, projects can fail to integrate or underdeliver.

Mitigation Strategies:

- Adopt a layered architecture like the 360 Smarter Stack, ensuring clarity of roles, boundaries, and dependencies between systems.

- Start with a Proof of Concept (PoC) in a live environment to validate key assumptions before full-scale rollout.

- Insist on open protocols and non-proprietary integrations to reduce vendor lock-in and ensure long-term adaptability.

- Run independent Technical Design Assurance reviews at key milestones to de-risk architectural decisions.

2. Operational Risk Mitigation

A common failure point in smart office projects is the gap between project completion and operational readiness. The most elegant system won't succeed if facilities, IT, or support teams aren't equipped to manage it.

Mitigation Strategies:

- Embed FM and IT stakeholders early in the design process, ensuring alignment on SLAs, escalation paths, and training needs.

- Define and document support models — who owns what, how alerts are triaged, and how issues are resolved.

- Schedule a structured handover phase post-installation, including documentation, training sessions, and tuning support.

- Use automation and monitoring tools to reduce manual oversight and empower proactive issue resolution.

3. Financial Risk Mitigation

Cost overruns, delayed benefits, or underestimated effort can undermine the value of a smart office initiative. These are among the most visible risks for decision-makers.

Mitigation Strategies:

- Develop a phased investment model, starting with core use cases and proving value early before expanding scope.

- Include change management and contingency allowances in the initial budget — avoid pretending that scope won't shift.
- Use fixed-price consultancy where possible, particularly for design or strategy phases.

- Track benefits realisation post-deployment and adjust expectations transparently.

4. Security & Data Privacy Risk Mitigation

Smart offices generate a wealth of sensitive data — including occupancy trends, access logs, and behaviour patterns. Mishandling this data could result in reputational damage or legal consequences.

Mitigation Strategies:

- Adopt privacy-by-design principles, anonymising data where possible and ensuring user transparency.

- Ensure all IoT and platform vendors meet IT security standards, such as ISO 27001, and undergo cybersecurity reviews.

- Collaborate with your InfoSec team early, so security isn't a blocker late in the project.

- Ensure data governance policies are clear — who owns data, who accesses it, and how long it's retained.

5. Change Management Risk Mitigation

Perhaps the most underestimated risk is people. Without proper engagement, even the best-designed smart office can struggle with adoption.

Mitigation Strategies:

- Conduct stakeholder mapping to understand who needs to be consulted, informed, and trained.

- Run tailored communications campaigns that explain the "why" behind new tools or processes.

- Pilot key features with small groups and incorporate feedback loops.

- Ensure senior leadership visibility and alignment, which can model behaviour and boost uptake.

Making Risk Mitigation a Visible Asset

Rather than hiding risks, the most compelling business cases highlight them alongside mitigation strategies. This:

- Builds trust with decision-makers

- Shows maturity and realism

- Differentiates your project from "tech for tech's sake" proposals

By proactively addressing risks, you not only safeguard delivery — you demonstrate that smart transformation is not just exciting, but sustainable, scalable, and strategically grounded.

Long-Term Value vs. Short-Term Cost Pressures

In almost every smart office programme, there comes a critical moment of tension: when visionary aspirations meet the realities of financial governance. Despite the growing consensus that smart workplaces are essential to productivity, sustainability, and wellbeing, organisations often face internal friction between long-term strategic value and short-term capital limitations. It's a classic clash — between the need to invest for future gain, and the pressures to minimise spend in the current budget cycle.

This section unpacks that dynamic and equips you with the language, insight, and tools to confidently reframe cost discussions around value, resilience, and return on investment.

The Pitfall of "Cheapest Wins"

Procurement processes — particularly in large or public-sector organisations — often follow a rigid cost-minimisation logic. Tenders are weighted toward lowest initial cost, and solutions are selected based on immediate affordability rather than lifecycle impact.

This approach, while fiscally conservative on the surface, introduces major hidden risks:

- Short Product Lifecycles: Lower-cost AV, sensor, or network components may not meet enterprise durability standards. Replacing them within 2–3 years results in additional spend and operational disruption.

- Lack of Integration: "Off-the-shelf" or closed systems may not support open APIs or interoperable standards, limiting future scalability and automation.

- Poor User Adoption: Technology that feels clunky, unreliable, or unintuitive will likely be underused — reducing the impact of the investment and lowering employee satisfaction.

- Missed ESG & Operational Efficiencies: Cheaper systems may lack advanced data features or energy optimisation capabilities, meaning you're not unlocking the operational value your business case promised.

In short, lowest cost often equates to lowest long-term value.

From Cost-Centric to Value-Centric Thinking

To advocate for smarter investment, project leaders need to reframe procurement conversations from price tags to performance outcomes. This requires a shift in mindset from "how cheap can we make it?" to "how effective can we make it over time?"

Key points of this reframing include:

- Strategic Alignment: Does the proposed investment support our ESG, hybrid working, or digital transformation goals?

- Flexibility & Scalability: Will this infrastructure adapt as business needs change?

- User Impact: Will it improve productivity, reduce friction, and elevate the workplace experience?

- Supportability: Is the chosen system maintainable, updatable, and future-proof?

Without this wider lens, decisions risk being penny-wise, pound-foolish.

Total Cost of Ownership (TCO): The True Financial Picture

Smart office business cases should always be evaluated through a Total Cost of Ownership (TCO) lens. TCO takes into account not just what something costs today, but what it will require — and return — across its operational life.

TCO considerations include:

Cost Element	Description
Capital Expenditure (CapEx)	Design, hardware, installation, initial setup
Operating Expenditure (OpEx)	Licensing, cloud services, support contracts
Energy & Resource Use	Impact of smart systems on utility spend
Support & Maintenance	Helpdesk load, break/fix cycles, tech refresh
Training & Change Management	Getting users comfortable and productive
Downtime & Frustration Costs	Cost of lost productivity or resistance to change
Replacement Cycles	Likelihood and timing of upgrades/replacement

By including these dimensions, a solution that looks more expensive at the outset may turn out to be more economical, sustainable, and impactful than its lower-cost alternatives.

Demonstrating Long-Term Value

It's essential to quantify — wherever possible — the long-term value your project will deliver. This can include:

- Space Utilisation Gains: Smart sensors and booking systems often reduce unnecessary desk and meeting room allocation, saving floor space or deferring new leases.

- Energy Efficiency & ESG Tracking: Intelligent lighting, HVAC, and occupancy systems contribute to carbon footprint reduction, support sustainability audits, and help meet compliance requirements (e.g., SECR, GRESB).

- Improved Productivity & Retention: Environments that work seamlessly improve staff engagement, wellbeing, and retention — a massive hidden cost centre for many firms.

- Better Decision-Making: Real-time workplace analytics enable facilities and IT teams to respond quickly and plan more effectively, avoiding costly reworks or inefficient operations.

These are the kinds of benefits that compound over time — and far outweigh small savings made by buying the "cheapest" product.

Overcoming the Budget Wall: Practical Strategies

In the real world, you may still face a wall of cost sensitivity. Here are strategies that can help maintain momentum without losing sight of long-term value:

- Phased Rollouts: Deliver in modular tranches, showing incremental ROI and using early success to unlock further budget.

- Pilot to Prove: Implement small proof-of-concepts that showcase impact and justify scale-up.

- Blended Funding Models: Combine CapEx and OpEx where appropriate — e.g., through "as a service" models for AV, access control, or analytics platforms.

- Cross-Departmental Cost Sharing: Align your project with HR (wellbeing), Sustainability (ESG), and IT (security and data) to co-fund initiatives.

- Hard KPIs for Soft Outcomes: Convert qualitative benefits (e.g., staff wellbeing) into quantifiable metrics (e.g., reduced absenteeism, higher NPS).

In Conclusion

A layered smart office transformation is not a "cost centre" — it is a value multiplier. When you articulate the strategic upside, account for lifecycle costs, and demonstrate the real-world impact of intelligent investment, you shift the conversation from "how much does it cost?" to "how much does it enable?"

The best smart workplaces are not those that cut the most corners — they're the ones that create the most value, both now and in the years ahead.

Layer 2: Operations & Culture

Chapter 4: Readiness of FM, IT & Business Teams

Even the most innovative smart office solutions can fall flat without one critical ingredient: operational readiness. It's not just about deploying new systems; it's about ensuring that the teams responsible for maintaining, supporting, and evolving those systems are prepared — both technically and culturally.

As workplaces become increasingly tech-enabled, traditional boundaries between Facilities Management (FM), Information Technology (IT), and the wider business functions are dissolving. Smart systems blur the line between the physical and the digital — requiring collaborative ownership, shared skillsets, and a unified vision of service delivery.

This chapter explores what it takes to get these teams ready for the reality of running a smarter environment — including the right skills, support structures, and service frameworks to ensure long-term success.

Skills and Support Models for Operating Smart Systems

As offices evolve into dynamic, data-rich environments, the skillsets and operational models needed to manage them must evolve too. The traditional FM model — focused on mechanical systems, vendor callouts, and routine maintenance — is no longer sufficient. Smart offices demand digital dexterity, cross-functional collaboration, and proactive service thinking.

To run a smart environment effectively, teams must blend technical proficiency, data fluency, and a service-first mindset. Below, we unpack what this looks like in practice.

The Core Competencies of Smart Operations

1. Digital Systems Familiarity

- Staff must become comfortable managing a landscape that includes IoT devices, IP-based building systems, web-based dashboards, and cloud control panels.

- This includes understanding protocols like BACnet/IP, MQTT, or API-driven data flows between systems (e.g., HVAC talking to occupancy sensors).

- Teams should know how to navigate vendor portals, view system health, apply firmware updates, and read log data.

2. Data Interpretation and Analytics

- Smart offices generate streams of data — occupancy, energy consumption, air quality, desk/room usage — but without analysis, it's just noise.

- Facilities and workplace teams need training in basic data literacy: recognizing trends, filtering anomalies, and making data-led decisions (e.g., how to optimize cleaning schedules based on usage).

- Understanding KPIs like energy per sqm, room utilization %, or sensor uptime is vital to ongoing performance.

3. Software and App Management

- Teams may now support workplace apps for hotdesking, visitor registration, wayfinding, or feedback capture.

- Administering user accounts, monitoring uptime, understanding backend configurations, and liaising with vendors becomes part of BAU (Business As Usual).

4. Cybersecurity Awareness

- Smart building components are increasingly IP-connected, which opens them up to cyber threats.

- Teams need foundational awareness of secure practices — password policies, access control, patching schedules, and understanding potential vulnerabilities (e.g., exposed APIs).

- IT involvement must extend beyond traditional networks to support and secure OT (Operational Technology) systems.

5. Soft Skills & Service Orientation

- Staff must handle user-facing technologies — digital signage, room panels, app support — and respond in a timely, user-centric way.

- This means communication skills, troubleshooting etiquette, and service empathy are as important as technical know-how.

Evolving the Support Model

Smart environments thrive on integration — not just between technologies, but between people and processes. Legacy FM and IT support models often fail in smart contexts because they're reactive and siloed. Here's how to reframe support:

1. Tiered, Integrated Support Frameworks

Adopt a structured model:

- Tier 0: Self-service (FAQs, dashboards, how-to guides).

- Tier 1: Generalist Helpdesk (log and triage issues, manage simple resets or reboots).

- Tier 2: FM/IT Specialists (handle deeper tech issues, liaise with vendors).

- Tier 3: Vendor Escalation (software bugs, major failures, firmware issues).

Clearly define who owns what across the smarter stack — for example, who monitors sensor health vs. app performance?

2. Smart Operations Team or Centre of Excellence

- In larger organisations, this could be a new team sitting at the intersection of FM, IT, and workplace strategy.

- Their job: Own smart tech platforms, ensure governance, review data trends, lead vendor engagements, and drive user adoption.

3. Managed Services and Vendor Partnerships

- Not all teams can support complex building systems internally. Strategic outsourcing of certain functions — e.g., remote monitoring of BMS, sensor uptime, or analytics dashboards — is common.

- Clearly define SLAs and accountability when outsourcing smart system support.

Training, Upskilling, and Knowledge Transfer

No team is instantly "smart-ready." A phased capability-building approach is essential:

- Vendor-Embedded Training: Negotiate knowledge transfer as part of system delivery. Ensure ops teams are part of FAT/SAT (Factory/ Site Acceptance Testing).

- On-the-Job Learning: Shadow vendors and commissioning teams. Run internal 'tech days' to demystify systems.

- Playbooks and SOPs: Develop system-specific manuals, escalation flows, checklists, and 'day in the life' guides.

- Scenario Training: Use drills and simulations to prepare for outage, performance drops, or user escalations.

- Continuous Development: Encourage CPD via professional bodies (e.g., IWFM, BICSI, CIBSE) or vendor certification schemes.

Common Pitfalls to Avoid

- No Support Plan in Place: New systems go live but there's no clear support route — resulting in slow fixes, finger-pointing, or reliance on project teams post-handover.

- Unrealistic Expectations: Assuming legacy FM or IT teams can suddenly run complex smart systems without training or resources.

- System Silos: Each system (AV, lighting, access control) has its own vendor, portal, and support route — creating a fragmented experience and poor integration oversight.

- Reactive Culture: Waiting for problems to emerge rather than proactively monitoring performance metrics and alerts.

Checklist: Are You Ready to Operate Smart?

- Do FM and IT teams understand the full tech stack they will be supporting?

- Are support tiers, playbooks, and escalation routes documented?

- Has hands-on training occurred prior to go-live?

- Are internal stakeholders comfortable with the data and tools now available to them?

- Is there a defined process for continuous review and improvement?

By investing in capability, clarity, and culture, organisations position themselves not just to operate smart systems, but to extract their full value. A great smart system is one that fades into the background — quietly optimising experience and performance — thanks to a well-prepared team behind the scenes.

The Convergence of IT and Facilities

From Parallel Tracks to Unified Pathways

In most legacy organisations, Information Technology (IT) and Facilities Management (FM) have historically operated in siloed domains, guided by different priorities, budgets, systems, and even languages. IT focused on networks, servers, cybersecurity, and digital enablement. FM focused on physical comfort, operational uptime, compliance, and space utilisation.

Smart buildings — and by extension, smart offices — have fundamentally upended this separation. The rise of integrated, sensor-rich environments means that what was once physical now generates data, and what was once digital now relies on environmental context. From a building's occupancy patterns to real-time air quality metrics, from touchless entry systems to dynamic space booking — the modern workplace is increasingly a fusion of code and concrete.

This convergence is not just a technical alignment; it's an organisational transformation. It represents a strategic and operational shift in how businesses must think about space, service delivery, user experience, and support.

Why Smart Demands Convergence

A smart office is not a collection of point solutions; it's an orchestrated ecosystem. Lighting systems respond to occupancy data; HVAC is regulated by external temperature and internal CO_2 sensors; meeting rooms are booked via mobile apps; digital signage pulls data from content management systems and user calendars.

No single team can manage all this complexity alone:

- FM needs to understand IoT networks, software update cycles, and API integrations.

- IT needs to understand BMS protocols, uptime for life-safety systems, and regulatory compliance in the built environment.

The overlap is no longer incidental — it's essential.

New Competencies and Role Blending

In response, we're seeing the emergence of hybrid roles and cross-functional teams. Titles like:

- Head of Smart Environments

- Digital Facilities Lead

- Smart Infrastructure Architect

- Converged Operations Manager

…are becoming common in forward-thinking organisations. These individuals possess dual fluency in both technical and operational languages — able to converse in packets and pipes, but also in maintenance schedules and occupancy KPIs.

Organisations must invest in upskilling existing teams and hiring for convergence-ready roles — those with experience in AV over IP, sensor platforms, space analytics, cyber-physical security, and cloud-based facility platforms.

Creating a Converged Delivery and Operating Model

Achieving true convergence isn't just about hiring differently — it's about designing systems and workflows to support shared ownership:

- Unified design principles: Involving IT, FM, security, and workplace teams at the planning stage of any fit-out or refurbishment project.

- Joint delivery governance: Establishing steering groups and design authorities where responsibilities, escalation paths, and integration points are formally agreed.

- Collaborative procurement: Ensuring vendors are evaluated based on both FM performance metrics and IT security/integration standards.

- Shared data models: Where space utilisation, energy, and occupancy data flow into centralised dashboards or BI tools accessible across disciplines.

- Integrated support channels: With tiered escalation across IT helpdesks and FM callouts, particularly for overlapping systems like meeting room tech or workplace apps.

Navigating Security and Operational Complexity

One of the most overlooked challenges in convergence is cybersecurity.

Building systems were never traditionally designed with endpoint security in mind. Many use legacy protocols (e.g., BACnet, Modbus) or operate over unsecured networks. As these systems are increasingly connected to enterprise IT infrastructure — or even exposed to the internet — risk increases dramatically.

IT must work with FM to:

- Segment networks for OT (Operational Technology) devices.

- Apply vulnerability management and firmware patching policies.

- Deploy firewalls and monitoring tools specific to building systems.

- Develop incident response playbooks for breaches that impact building control.

At the same time, FM needs to understand how system configuration changes (e.g., open ports, cloud integrations) can introduce unintended risk.

Building a Culture of Collaboration

Technology alone won't solve the convergence challenge. It requires a deliberate cultural shift.

- FM teams need to move beyond reactive maintenance to data-driven service planning.

- IT teams must go beyond network uptime to experience-led delivery.

- Both teams must participate in joint training, attend cross-discipline workshops, and be part of shared OKRs and KPIs that reflect the success of smart initiatives.

Communication is key — not just about what each team does, but how their decisions impact others. An API schema change by IT might break a dashboard managed by FM. A firmware update by FM might interrupt a cloud sync needed by IT.

The Road Ahead: From Convergence to Co-ownership

In the long term, smart offices will necessitate co-owned systems and cross-functional leadership. We'll see:

- Digital twins that map not just space but system interdependencies, available to all departments.

- Enterprise command centres for building performance, bringing IT, FM, and business teams together.

- Smart service teams that blend engineers, developers, and analysts in shared squads.

- Common performance scorecards where IT and FM are evaluated on user experience, energy efficiency, sustainability outcomes, and technology adoption — not just ticket volumes or uptime.

True convergence is not just collaboration — it's integration. It's when Facilities and IT stop seeing themselves as separate service providers and begin operating as a single, smart workplace enablement function. That's when the full value of smart technology can be realised.

Aligning with SLAs, Service Design, and Culture Change

Why This Alignment Is Vital

As smart workplaces grow in complexity, their success no longer hinges on technology alone, but on how well that technology is integrated into daily operations, support models, and workplace culture. This means looking beyond installation and configuration to ask:

Who will maintain this? Who supports the end-user? How do we measure whether it's working? How do we evolve as user needs change?

To answer those questions, smart workplace delivery must be aligned with:

- SLAs (Service Level Agreements) that reflect the real performance expectations of smart systems

- Modern service design, which ensures smooth handoffs between tech, people, and processes

- A culture shift that embraces continuous learning, cross-functional collaboration, and technology as a service enabler

This alignment ensures that once a smart system goes live, it is supported, understood, trusted, and used, rather than becoming a forgotten or underutilised asset.

Reimagining SLAs in the Smart Era

Traditional SLAs were created in a time when building systems were mostly siloed and static — think HVAC uptime, lighting maintenance, or software availability measured in raw percentages. But smart workplaces depend on dynamic, data-driven systems where value isn't just uptime — it's performance plus relevance.

A sensor may be operational, but if it's providing stale or inaccurate data, the system fails. A room booking panel may be online, but if it's not synced with occupancy sensors, user trust collapses. Therefore, smart SLAs must account for real-world usability and effectiveness, not just availability.

Examples of new SLA dimensions include:

- Accuracy guarantees for sensor data (e.g., <5% variance from manual spot-checks)

- Latency thresholds for data updates in dashboards or analytics

- Integration SLAs to measure success of API-based interoperability

- Issue triage response times for cross-domain issues (e.g., IT/FM overlap)

- UX-specific SLAs, like resolving digital signage outages within X hours

You're no longer just promising uptime — you're promising insight, reliability, and actionable performance. These new SLAs must be owned jointly by IT, FM, and any third-party service providers under a shared delivery model.

Service Design That Works at the Speed of Smart

The shift from static to smart also demands a rethink in service design — the end-to-end model for how services are delivered, supported, escalated, and evolved.

The Smart Office Project

In a smart office:

- Devices are distributed across the workplace (from door sensors to edge gateways)

- Systems are interconnected (e.g., access control triggering lighting and HVAC sequences)

- User expectations are higher, shaped by the convenience of consumer tech

This means services must be designed around:

- Multi-layer ownership: clear delineation of who supports physical devices, software, networks, cloud platforms, and user-facing experiences

- Proactive monitoring and diagnostics, not just reactive break/fix

- Real-time service transparency for workplace teams and end-users (e.g., knowing when a sensor last updated or when a room booking issue is being investigated)

- Integrated ticketing across FM and IT workflows — ideally with smart categorisation and auto-routing for efficiency
- Digital twins and asset intelligence to reduce mean time to repair (MTTR) and improve first-time fix rates

A well-designed service model enables faster support, fewer escalations, and more confident end-users, all while scaling with minimal manual effort.

Embedding a Culture That Embraces Change

The most advanced smart infrastructure will still underdeliver if people aren't on board, engaged, and empowered. This is why culture change is not an afterthought — it's a core pillar of smart workplace success.

- Smart environments often challenge existing ways of working:

- New ways of booking space, accessing areas, or reporting issues

- Data-led decision-making in areas previously driven by gut feeling

- Transparency about how space and services are used — sometimes raising privacy or surveillance concerns

- More self-service and digital engagement — which can be unfamiliar or intimidating for some users

To navigate this, organisations need a culture of digital confidence, agility, and cross-functional collaboration. That means:

- Smart champions embedded across business units to advocate for and demystify the technology

- Ongoing training and onboarding — not just for launch, but as systems evolve

- Feedback mechanisms (e.g., in-app surveys, analytics dashboards, town halls) to keep systems responsive and user-centric

- Visible leadership support, with execs modelling the use of smart systems and reinforcing digital behaviours

- A culture of experimentation, where teams feel safe to try new ways of working and refine them quickly based on evidence

This cultural maturity transforms smart systems from tools into trusted companions in the workplace experience.

Alignment in Action: A Continuous Loop

When SLAs, service design, and workplace culture are aligned:

- Issues are caught earlier, because monitoring is intelligent and shared across domains

- Users report fewer problems, because systems are reliable, intuitive, and supported

- Support teams spend less time firefighting, and more time improving

- Stakeholders feel ROI, not just in tech, but in engagement, resilience, and brand perception

Most importantly, it creates a continuous improvement loop. User feedback informs service adjustments. Analytics drive SLA refinement. Culture evolves to support more advanced capabilities — which in turn fuels greater outcomes.

Chapter 5: Governance & Change Management

Implementing a smart office is far more than a technical deployment — it's an organisational transformation. As systems, teams, and services converge, maintaining clarity, accountability, and alignment becomes mission-critical. Without structured governance and proactive change management, even the most promising smart initiatives risk stalling due to ambiguity, resistance, or fragmented ownership.

This chapter explores how to establish the right oversight and governance mechanisms — such as steering groups, structured decision-making models, and clear escalation paths — alongside a robust strategy for training, adoption, and behavioural change. Together, these provide the scaffolding to move from pilot to scaled success.

Steering Groups and Decision-Making Frameworks

Why Smart Workplaces Need Smarter Governance

Smart workplace projects often span a wide mix of disciplines — from real estate, IT, and FM to HR, legal, and sustainability — all with different priorities, vocabularies, and success measures. Without a formalised governance structure, decisions can become siloed, duplicated, or delayed. That's why a clear and empowered steering group becomes not just helpful, but essential.

This group becomes the central nervous system of the programme — absorbing information, coordinating direction, and enabling timely, aligned decisions across diverse functions.

Structure and Composition: The Right People at the Right Table

The composition of the steering group must reflect the 360-degree nature of smart office delivery. Think of it as building an alliance between strategic and operational leaders, ensuring no critical lens is overlooked.

Typical members might include:

Role	Contribution
Facilities / Workplace / RE Lead	Operational readiness, vendor coordination, lifecycle support
IT / Digital Transformation	Infrastructure compatibility, data security, integrations
HR / People Experience	Workplace culture, wellbeing, hybrid policy alignment
Business Leadership	Strategic objectives, department-specific needs, funding
Legal / Compliance	Data protection, privacy, contracts, regulatory standards
Finance / Procurement	Cost assurance, ROI validation, procurement oversight
Change Manager / Communications	Behavioural change, user engagement, adoption metrics
Sustainability Officer	ESG alignment, energy tracking, environmental impact
External Consultants / Specialists	Independent validation, technical insight, vendor neutrality

Tip: Include end-user reps (e.g. EA, team manager) where possible. Their perspective brings real-world insight into how new systems land on the ground.

Functions of the Steering Group

The steering group should operate with clear purpose and discipline. While the exact remit will vary by organisation, its key functions typically include:

- Strategic Oversight: Ensuring alignment with business objectives, ESG goals, and digital transformation strategies.

- Prioritisation and Scope Control: Managing what gets delivered first and what may be deferred or dropped.

- Budget and Resource Monitoring: Ensuring allocation aligns with value and delivery milestones.

- Risk Management: Identifying interdependencies, compliance issues, and technical or people-based risks.

- Stakeholder Communication: Coordinating consistent messaging and updates across the wider business.

- Decision Acceleration: Providing clear approval mechanisms to prevent bottlenecks.

Frameworks That Enable Fast, Informed Decisions

A steering group must be more than a discussion forum — it must enable action. This requires structured frameworks that balance agility with control.

Here are a few effective approaches:

☐ RACI Matrices

Define who is Responsible, Accountable, Consulted, and Informed for every core component across the 360 Smarter Stack. This prevents duplication or blind spots, especially during system integration or rollout stages.

☐ Milestone Gate Reviews

Define key checkpoints where the steering group must approve progression (e.g. end of discovery, vendor selection, pre-go-live). These gates align technical delivery with business readiness.

☐ Decision Logs

Document what decisions were made, why, and by whom — a vital audit trail, especially as teams or leadership change over time.

☐ Weighted Scoring Models

Use defined scoring rubrics to evaluate smart workplace solutions — not just on cost, but also UX, sustainability, support model fit, scalability, etc. This makes vendor decisions transparent and defensible.

☐ Escalation Paths

If issues arise that threaten timelines, budget, or outcomes, there must be a fast and agreed path for escalation and resolution.

Meeting Cadence & Governance Rhythm

- Monthly Steering Group Meetings – For strategic updates, key decisions, risk management, and milestone reviews.

- Weekly Working Group Meetings – Tactical updates with project managers and technical leads.

- Quarterly Business Reviews (QBRs) – Zooming out to review outcomes, alignment, lessons learned, and roadmap revalidation.

Note: Avoid over-governance. The goal is not more meetings, but better decisions and forward motion.

Making Governance an Enabler, Not a Barrier

Too often, governance is perceived as bureaucracy. In a smart office context, it must be reframed as an enabler of innovation and accountability. By creating a structured but responsive steering model, teams are empowered to take action while still operating within guardrails that protect the business, users, and long-term success.

In summary, effective steering groups and decision frameworks:

- Build confidence across departments and leadership

- Prevent scope drift and costly misalignments

- Create visibility and transparency in progress and decisions

- Enable fast course correction when things change — as they inevitably do

Training, Adoption, and Embedding New Ways of Working

Training, Adoption, and Embedding New Ways of Working

In any smart workplace transformation, technology may be the most visible element — but people are the real lever of success. The most intelligent systems and beautifully integrated platforms will fall short if the workforce doesn't understand, accept, or embrace them. That's why investment in training, adoption, and cultural embedding is not a postscript — it's a core pillar of smart transformation.

This is especially true in organisations adopting the 360 Smarter Stack methodology. The Stack provides a clear, layered roadmap — and ensuring people at all levels understand how their role fits into this ecosystem helps drive shared ownership and alignment. Training and engagement must not only teach users how to use tools, but also why those tools matter within the wider transformation journey.

1. Training: Purposeful, Layered and Inclusive

Effective training is designed around people, not products. It must be inclusive, adaptable, and relevant. That means:

- Role-based training paths: Executives, support staff, FM teams, and IT each use the smart environment differently. Training should reflect those unique interactions, needs, and priorities.

- Progressive learning stages: Introduce key functions early and build deeper capability as familiarity grows. This helps prevent cognitive overload and promotes retention.

- Multi-format delivery: Blend digital tutorials, in-person sessions, mobile-friendly materials, and just-in-time learning embedded in the tools themselves.

- Peer support through champions: Identify early adopters or departmental leads as "Smart Ambassadors" who can support their teams, reinforce best practices, and act as feedback channels.

Where the 360 Smarter Stack adds further value is in providing a framework for layered training. For instance, one team may focus on user experience (Layer 3), another on integration (Layer 6), while senior stakeholders align with strategic outcomes (Layer 1). This structured awareness supports cross-functional understanding, helping to break down silos and create stronger delivery momentum.

2. Adoption: Driving Behavioural Change, Not Just Usage

Smart technology must do more than function — it must change how people work for the better. Adoption means moving beyond awareness and usage to create confident, value-driven behaviours:

- Encouraging users to trust data-driven decisions, such as real-time occupancy info for planning.

- Supporting self-service use of workplace apps to book, report, collaborate, or access analytics without friction.

- Replacing outdated processes with new workflows that reflect the smart capabilities (e.g., auto-check-ins or integrated fault reporting).

Adoption is most successful when change feels intuitive, relevant, and beneficial. This can be achieved by:

- Mapping user journeys and optimising pain points in everyday interactions.

- Creating meaningful adoption metrics (such as reduction in manual interventions or improved service response).

- Piloting with real users to shape the final experience before full-scale rollout.

Because the 360 Smarter Stack is designed to be iterative and scalable, adoption efforts should mirror that. Layer by layer, teams should grow in capability and confidence — whether it's engaging with building data (Layer 5) or managing digital signage and wayfinding (Layer 4).

3. Embedding Smart Ways of Working into Organisational Culture

Training and adoption are the starting points — but long-term success hinges on embedding change into culture.

This involves:

- Visible leadership: Senior figures consistently using and championing the smart systems demonstrates commitment and drives wider engagement.

- Policy alignment: HR, IT, and FM policies must evolve to reflect new realities — such as hybrid meeting norms, app-based reporting, or AI-powered usage insights.

- Ongoing communication and storytelling: Share successes, user stories, and performance improvements. Keep the narrative alive beyond go-live.

Additionally, organisations must align incentives and link smart engagement to business KPIs. For example, if a goal is energy efficiency, make performance dashboards visible and team-based energy challenges part of the culture.

Finally, the 360 Smarter Stack encourages continuous feedback and refinement. Embedding smart ways of working is not a one-off activity — it's a living cycle. By using the Stack to assess maturity across each layer, teams can prioritise new areas of focus and ensure the environment continues to evolve with user needs.

In summary, when people are trained with clarity, when adoption is supported with purpose, and when new ways of working are reinforced at all levels, smart workplace investments begin to pay off in lasting, measurable ways. The 360 Smarter Stack provides the scaffolding for this transformation — but it's people who bring it to life.

Layer 3: User Experience

Chapter 6: Designing for Humans

As much as smart technology relies on infrastructure, data, and automation, its ultimate purpose is to enhance the experience of people — the employees, visitors, contractors, and service staff who interact with the workplace every day. A successful smart office doesn't just work; it works for people.

Designing for humans means stepping into the shoes of those users — understanding their needs, frustrations, goals, and expectations at every touchpoint. It's about crafting journeys that are intuitive, efficient, and even delightful. Smart offices should remove friction, not add complexity.

To achieve this, we need to blend UX thinking with spatial design, underpinned by the 360 Smarter Stack methodology — particularly focusing on Layer 3: User Experience and how it connects to the other layers above and below it. When user-centred thinking is embedded from the start, technology becomes invisible, and the workplace becomes seamless.

Let's begin by exploring one of the most vital tools in human-centric smart office design: mapping the user journey.

Mapping User Journeys

Mapping user journeys is a foundational technique in experience design, but in the context of smart office transformation, it becomes a strategic necessity. A smart office is only truly "smart" when it responds to the needs and behaviours of its users — employees, visitors, support teams, and service providers alike. By visualising and analysing these journeys, organisations can ensure the technology, spaces, and services they implement are truly human-centric.

Journey mapping allows us to shift perspective — from systems and infrastructure to the real, lived experience of people interacting with the workplace every day. It answers a core question: What does a day in the life of our users feel like in this environment?

Purpose and Outcomes

A well-constructed user journey map provides insight into:

- How people navigate the workplace — physically and digitally

- Which tools or services they interact with — and when

- What emotions and frustrations they encounter — and why

- Where automation, data, or technology can streamline or enhance their experience

- Which layers of the 360 Smarter Stack are actively supporting their journey

This approach is a key part of Layer 3: User Experience in the 360 Smarter Stack, but it also creates vital input for:

- Layer 4: Applications & Services — defining the functionality required

- Layer 2: Operations & Culture — surfacing training and support needs

- Layer 5: Data & Context — identifying key data points from user behaviour

Personas and Audience Types

Start by identifying your core user groups. These are typically represented by personas, each with distinct goals, access needs, and frustrations:

- Hybrid workers – split between home and office; seek ease of booking, space availability, and productivity tools

- Visitors and clients – value wayfinding, security, and a polished first impression

- Facilities and IT teams – need oversight, monitoring tools, and streamlined service workflows

- Executives/VIPs – expect frictionless experience, reliability, and rapid response

- Support and service staff – rely on consistent access, support requests, and mobility

Mapping journeys for each persona helps identify where shared experiences overlap and where tailored functionality is required.

The Five Key Journey Phases

To structure your map, break down the workplace experience into five main phases:

1. Pre-Arrival

- What users are doing: Checking room availability, booking desks or parking, reviewing who's in that day.

- Touchpoints: Mobile apps, calendar integration, chatbots, digital signage.

- Pain points: Inaccurate availability, unclear instructions, no mobile access.

- Opportunities: Predictive suggestions (AI), digital access passes, sustainability nudges (e.g., low-traffic days).

2. Arrival and Entry

- What users are doing: Entering the building, navigating to spaces, locating colleagues.

- Touchpoints: Access control systems, kiosks, signage, mobile wayfinding, reception areas.

- Pain points: Long queues, confusing navigation, unwelcoming first impression.

- Opportunities: Touchless entry, smart visitor check-in, digital concierge.

The Smart Office Project

3. In-Building Engagement

- What users are doing: Collaborating, using meeting rooms, taking breaks, adjusting comfort settings.

- Touchpoints: Room booking panels, collaboration tools, personal control apps, sensors, AV systems.

- Pain points: Double-booked rooms, inaccessible equipment, discomfort.

- Opportunities: Personalised preferences, environmental sensors, dynamic space allocation.

4. Support & Incident Management

- What users are doing: Reporting issues, tracking service status, receiving help.

- Touchpoints: Apps, kiosks, QR codes on equipment, helpdesks.

- Pain points: Long response times, unclear status, inconsistent support.

- Opportunities: Self-service tools, automated updates, AI-driven triage.

5. Departure and Follow-up

- What users are doing: Ending bookings, checking out, reflecting on experience.

- Touchpoints: Exit sensors, feedback surveys, app notifications.

- Pain points: Forgetting to check out, lack of closure or follow-up.

- Opportunities: Usage summaries, wellness reports, feedback loops for service improvement.

Journey Mapping in Practice

Journey mapping can be executed via:

- Workshops with cross-functional teams (IT, FM, HR, end-users)

- Digital tools like Miro, FigJam, or Service Blueprint templates

- Empathy mapping, to understand user emotions alongside actions

- Shadowing or observation, to gain real behavioural insights

In these sessions, plot each step along the user's journey, annotate with emotions, pain points, touchpoints, and technologies involved, and connect those to relevant Stack layers. This ensures you're not just digitising experiences — you're strategically aligning them with the overall delivery architecture.

Using Journey Maps to Drive Design

Once mapped, journey insights feed directly into:

- Design briefs for UX, signage, app developers, and space planners

- Requirements documents for technology procurement and integration

- Training programmes tailored to actual user behaviour

- Change management plans that empathise with real needs

- KPI tracking that measures satisfaction and friction points

Ultimately, journey mapping provides the human narrative behind technical and operational decisions — and it's the anchor point for designing spaces and systems that people genuinely enjoy using.

Defining Physical and Digital Touchpoints

In a smart workplace, physical and digital touchpoints serve as the critical points of interaction between people and the environment. These touchpoints form the visible and tangible aspects of the smart office experience — influencing how users navigate spaces, access services, personalise their environment, and feel supported in their daily work.

Without well-designed and thoughtfully integrated touchpoints, even the most advanced smart systems risk becoming invisible or underutilised. Touchpoints are how technology becomes human — and it's in their design, placement, and functionality that user adoption, satisfaction, and performance are ultimately won or lost.

Importantly, every touchpoint must be anchored in a broader delivery strategy, aligned with the 360 Smarter Stack methodology, particularly Layer 3 (User Experience), Layer 4 (Applications & Services), Layer 6 (Integration Layer), and Layer 7 (Systems & Devices).

Physical Touchpoints – Real World Meets Real-Time

Physical touchpoints are the tangible, often shared interfaces people use throughout the workplace. They need to be intuitive, durable, and context-aware — and where appropriate, they should feel seamless rather than "techy," blending into the fabric of the space while delivering meaningful value.

Examples include:

◆ Digital Signage & Wayfinding Displays

- Positioned in lobbies, lift lobbies, or corridors to inform staff and visitors about real-time space availability, events, travel disruptions, or sustainability metrics.

- Dynamic content updates (powered by live data) can reflect occupancy, air quality, or emergency alerts.

- Touch-enabled signage can offer interactive wayfinding or directory search in larger campuses.

◆ Room Booking Panels

- Mounted outside meeting rooms or enclosed spaces to show real-time occupancy status, future bookings, and offer quick reservation functionality.

- Panels with NFC or QR integration allow fast check-in/check-out.

- Integration with workplace calendars, occupancy sensors, and building controls creates a more fluid experience — for instance, releasing rooms when no one checks in.

◆ Visitor Kiosks & Access Points

- Self-service kiosks allow visitors to check in, print badges, notify hosts, and receive wayfinding support.

- Facial recognition or mobile pass handoffs can streamline secure entry and minimise touch.

◆ Interactive Environmental Controls

- Wall-mounted or desktop interfaces that allow employees to adjust lighting, blinds, or temperature — typically within controlled parameters.

- When linked to occupancy sensors and user profiles, the system can auto-adjust conditions as people arrive at specific desks or zones.

◆ Smart Furniture & Embedded Touchpoints

- Meeting tables with integrated touchpads, wireless charging, or booking displays.

- Lockers or sit/stand desks linked to mobile apps for personalisation or usage tracking.

Digital Touchpoints – Personalised, Portable, Predictive

Digital touchpoints complement the physical by offering anytime, anywhere access to services, controls, and insights — making the workplace experience more user-centric, mobile, and on-demand.

These touchpoints play a vital role in enabling hybrid work, extending office capabilities beyond the building, and supporting continuous engagement.

◆ Mobile Workplace Apps

- Offer centralised control over desk booking, meeting room search, wayfinding, issue reporting, catering orders, and more.

- Integrate with personal calendars, access control, building automation, and IoT systems to create a truly smart experience.

- Push contextual notifications such as room readiness, delays, maintenance alerts, or air quality issues.

◆ Chatbots & Digital Assistants

- Embedded in platforms like Teams, Slack, or web portals to offer conversational access to services.

- Can help users find available spaces, reset passwords, report problems, or request support.

◆ Personal Control Dashboards

- Accessible via app or web, these dashboards allow users to personalise lighting, temperature, noise preferences, or book their preferred desk setup.

- Dashboards can display real-time metrics such as indoor air quality, noise levels, or cleaning schedules.

◆ Touchless Check-in via QR/NFC

- Smart signs, desk placards, or room displays can allow users to check in with a tap or scan, eliminating friction and improving utilisation data accuracy.

◆ Intranet & Workplace Portals

- Often integrated with the organisation's digital ecosystem, portals offer a self-service hub for workplace policies, building guides, IT support, and employee services.

Multimodal Experiences: Bridging the Physical and Digital

The most effective smart office environments don't separate physical and digital touchpoints — they weave them together into coherent, responsive journeys.

Examples include:

- A visitor receives a QR-based access pass before arrival → uses it to enter via a secure turnstile → is guided to a meeting room by digital signage → checks in at the room panel → receives a prompt for feedback via their mobile app post-meeting.

- An employee books a desk from their app the night before → arrives to find the desk lit green with climate preset to their preference → uses their phone to control nearby lighting or raise a facilities ticket.

- This orchestration depends on cross-layer integration — Layer 4 (Applications), Layer 5 (Data), Layer 6 (Integration), and Layer 7 (Devices) all working together to serve Layer 3: User Experience.

Design Considerations for Touchpoints

To ensure each touchpoint adds value and doesn't become digital clutter, teams should evaluate:

- Frictionless Access: Can users get what they need in just 1–2 steps?

- Consistency: Are symbols, colours, and interaction styles standardised across touchpoints?

- Accessibility & Inclusivity: Can people of all abilities use them easily?

- Context Awareness: Does the interface adapt to location, time, role, or usage history?

- Privacy & Security: Are personal data and interactions handled appropriately?

- Training & Support: Do users know what the touchpoint does and how to use it?

Every interaction matters. When thoughtfully integrated, physical and digital touchpoints become enablers of efficiency, wellbeing, and seamless experience. They turn abstract concepts like "smart workplace" into something tangible, usable, and human.

And when planned with the 360 Smarter Stack, organisations can ensure that each touchpoint is not a siloed gimmick, but a part of a carefully layered and interconnected delivery model — enabling consistent, scalable, and meaningful transformation.

Accessibility, Inclusivity, and Experience Design

In the pursuit of smarter, tech-enabled workplaces, there's a vital truth we must not overlook - technology should serve everyone. If we fail to account for how people experience the workplace, we risk building spaces that are efficient but alienating — intelligent but inaccessible.

This is why accessibility, inclusivity, and experience design must be treated as foundational pillars of any smart office initiative, not as optional extras or compliance tick-boxes. These principles are not only about meeting legal requirements — they are about enabling dignity, belonging, and effectiveness for every individual who interacts with the environment.

The 360 Smarter Stack places these considerations front and centre. From strategic vision (Layer 1) to UX and engagement (Layer 3), applications (Layer 4), and devices and infrastructure (Layers 7 and 8), the model demands a user-first mindset at every stage. In doing so, it ensures that accessibility and inclusivity aren't just surface features — they are embedded into the DNA of the modern workplace.

Accessibility: From Obligation to Opportunity

Accessibility is often framed through the lens of obligation — as something organisations must do to comply with disability legislation. But when we shift perspective and treat accessibility as an opportunity — to reach more users, enable more ways of working, and create better design for all — the conversation changes.

An accessible smart office considers a wide range of user needs, including:

- Physical mobility: Step-free access, automated doors, wheelchair-friendly desk heights, and adjustable meeting room controls.

- Sensory impairments: High-contrast displays, screen readers, tactile indicators on signage, and real-time captioning in AV systems.

- Cognitive and neurodivergent needs: Simplified wayfinding, noise-controlled spaces, and interfaces that reduce information overload.

- Situational challenges: Accommodating temporary injuries, ageing users, or multitasking employees — all of whom benefit from intuitive, low-friction design.

Importantly, smart systems enable dynamic accessibility. Lights can automatically adjust to visual comfort levels. Voice-activated systems can replace touch-based interfaces. Digital signage can switch modes based on user proximity or preference. IoT sensors and AI can adapt the space to user needs in real time, breaking down traditional barriers to participation.

This is the promise of smart workplace accessibility: not just to accommodate the few, but to enhance the experience for everyone.

Inclusivity: Designing for a Diverse Workforce

Inclusivity expands on accessibility by asking a deeper question: Does everyone feel they belong here?

A truly inclusive smart office acknowledges that today's workforce is more diverse than ever — not just in terms of age, ability, and background, but also in workstyles, values, and cultural expectations. It recognises that the "average user" doesn't exist — and instead strives to serve a spectrum of experiences.

Inclusive smart office design means:

- Multiple modes of interaction: Touchscreens, mobile apps, voice interfaces, and analogue backups all coexisting to accommodate different comfort levels and preferences.

- Variety of spaces: Some people thrive in open, collaborative zones; others need quiet focus booths, prayer rooms, or recharge spaces. Smart zoning and room booking systems help ensure all are accessible.

- Cultural and linguistic sensitivity: Multilingual signage, diverse imagery in digital interfaces, and culturally neutral design all support a global workforce.

- Support for hybrid teams: Tools that bridge the digital/physical divide ensure remote participants can interact equally with on-site colleagues — from AV setups to digital whiteboards and mobile-first tools.

The goal isn't just parity — it's empowerment. Inclusivity means enabling every employee to feel seen, valued, and fully capable of engaging with their environment.

And the journey to inclusivity must be participatory: employee networks, diversity groups, and real user feedback loops should be involved early and often in smart office planning. The smartest decisions often emerge when we co-design with the people we're designing for.

Experience Design: The Emotional Layer of Smart Workplaces

Experience design is where the tangible and the intangible meet. It's not just about how things function — it's about how they feel. When employees enter a workplace, their first impressions, their ease of navigation, their interactions with technology — all of this affects morale, productivity, and wellbeing.

Smart workplace experience design focuses on:

- Flow and intuitiveness: Can users instinctively navigate space, book resources, and control their environment without instruction?

- Sensory balance: Lighting, acoustics, temperature, and visual stimuli all contribute to a sense of comfort or overstimulation. Smart systems can dynamically balance these based on occupancy or user input.

- Emotional resonance: Design choices — from interface tone to environmental branding — should reflect the organisation's culture and make people feel welcome, energised, or at ease.

- Feedback and learning: Systems should provide clear feedback ("Your desk is reserved"), anticipate needs ("You're near your next meeting room"), and offer help ("Would you like accessibility mode?").

The most powerful experience design makes the technology invisible — it just works, so that users can focus on doing their work.

And because experiences differ widely, smart design must be flexible and responsive. A system that allows personalisation — of lighting, temperature, desk height, or software dashboard — not only improves comfort but sends a clear message: you matter here.

Weaving It All Together: The 360 Smarter Stack Approach

Accessibility, inclusivity, and experience design aren't isolated considerations — they touch every layer of the 360 Smarter Stack, including:

- Layer 1 (Strategy & Vision): Inclusive design goals should be baked into business objectives and defined in success metrics.

- Layer 2 (Service & Support): Support teams must be trained to assist diverse users and ensure accessibility needs are promptly addressed.

- Layer 3 (UX & Engagement): User journeys must reflect the full spectrum of real-world users — not just the tech-savvy or able-bodied.

- Layer 4–7 (Apps, Integration, Platforms, Devices): All tech must follow inclusive design standards and support assistive technologies.

- Layer 8 (Infrastructure): Physical design and systems should enable equitable access and intuitive navigation.

Only by thinking holistically across all 8 layers can organisations ensure that their smart workplace investments serve the broadest possible set of users — and generate the strongest long-term returns.

Smart workplace transformation is about more than sensors, dashboards, and digital tools. It's about people — and people are wonderfully diverse.

When accessibility, inclusivity, and experience design are treated as strategic pillars — not secondary considerations — the result is an office that works better for everyone. It's not just smarter. It's fairer, more humane, and more effective.

The 360 Smarter Stack provides a systematic approach to embedding these values throughout the smart office lifecycle — making sure that no user is left behind, and every voice is reflected in the fabric of the workplace.

Chapter 7: Workplace Enablement

As the modern workplace evolves beyond fixed seating plans and static floor layouts, enablement has become a critical lever for smart transformation. It's not just about deploying new tools — it's about ensuring that people can use the environment fluidly, effectively, and with as little friction as possible.

Workplace enablement refers to the digital and physical systems that empower employees to navigate, interact with, and personalise their work environment. Whether that's booking a desk near a colleague, finding a quiet room for focused work, or seamlessly transitioning between home and office — smart technology must remove barriers, not introduce them.

The 360 Smarter Stack treats workplace enablement as a cross-layered consideration. It draws from user experience (Layer 3), workplace apps and services (Layer 4), data platforms (Layer 5), and physical systems (Layers 7 & 8). The result is a connected environment that actively supports how people want to work — not how buildings used to function.

Let's explore some of the key components.

Smart Desks, Booking Systems, Wayfinding

Workplace enablement is no longer a luxury — it's a foundational requirement. As organisations shift toward hybrid work models and demand greater agility from their spaces, the need for responsive, intelligent, and user-centric workplace systems has never been greater. Three core enablers stand out: smart desks, dynamic booking systems, and intelligent wayfinding. Each serves a unique function, but when integrated effectively, they create a seamless experience that transforms the office from a static space into an interactive platform for productivity.

Smart Desks – From Fixed Assets to Adaptive Resources

In the past, desks were assigned by nameplates and remained unchanging. Today's smart desk, however, is a dynamic workplace resource designed to respond to real-time needs and user preferences. These desks incorporate technology that allows for:

- Occupancy Detection: Embedded sensors detect whether a desk is occupied or available, feeding real-time data into floor maps and booking systems.

- Touchless Check-In: Employees can check into a desk via RFID badge, mobile app, or even proximity detection using Bluetooth or Wi-Fi.

- User Preferences: Profiles can be linked to height-adjustable desks, lighting preferences, or ergonomic reminders, creating a personalised experience.

- Utilisation Analytics: Facilities teams can track occupancy trends, average session durations, and high-demand zones — informing long-term space planning and desk-to-employee ratios.

- Environment Feedback: Integration with air quality, temperature, and lighting sensors allows users to see how their workspace is performing — with potential nudges to move to a better area.

Smart desks also form the link between human behaviour and digital insights, helping organisations evolve layouts, manage demand, and design better environments.

Booking Systems – Orchestrating Time, Space, and Resources

Booking systems are the nervous system of a flexible workplace. They give users the power to reserve not just desks, but also rooms, parking bays, lockers, focus pods, quiet zones, catering, or visitor passes — all from a single interface.

Modern booking systems offer:

- Context-Aware Booking: They suggest nearby desks based on meetings, recent bookings, or team locations. Some even integrate with AI assistants to recommend bookings aligned with your calendar.

- Rules & Permissions: Administrators can enforce policies like team-based neighbourhoods, alternating team days, or max occupancy limits — ensuring safe and structured use.

- Calendar Integration: Full interoperability with tools like Outlook, Google Calendar, or Microsoft Teams, so users don't have to switch platforms.

- Real-Time Status: Live availability is shown via desktop maps, mobile apps, digital signage, or even voice assistants. Users can see which areas are quiet, crowded, or booked.

- Sustainability Nudges: Platforms may suggest quieter or more energy-efficient zones — nudging users toward lower-carbon choices.

Critically, booking systems must orchestrate the entire workplace — tying together user preferences, space allocation, and business policies, while ensuring a consistent user experience across devices and platforms.

Wayfinding – Navigating Complexity with Confidence

In larger or unfamiliar office environments — especially post-pandemic where workers may attend different hubs — intelligent wayfinding helps users orient themselves and find their way efficiently.

Modern wayfinding includes:

- Interactive Digital Signage: Kiosks at entry points or lifts showing live maps, desk locations, and room statuses.

- Mobile Navigation: Step-by-step directions via smartphones or wearables, including indoor positioning with Bluetooth beacons, Wi-Fi triangulation, or smart lighting systems.

- AR Overlays: Emerging technologies allow users to hold up a phone and see directional arrows or desk identifiers layered onto their camera view.

- Accessibility Routing: Wayfinding systems should suggest routes for wheelchair users or individuals needing step-free access — as well as audio guidance for the visually impaired.

- Dynamic Updates: Real-time redirection if a lift is out of service, a route is congested, or a booked space becomes unavailable — mirroring the fluidity of transport apps like Google Maps.

Beyond simple navigation, smart wayfinding improves first impressions for visitors, speeds up the onboarding of new hires, and helps hybrid staff feel connected to their ever-evolving workspace.

Enabling Integration through the 360 Smarter Stack

All three of these enablers — desks, booking, and navigation — are not stand-alone solutions. They must work together as a connected ecosystem, which is precisely why the 360 Smarter Stack is so valuable. Here's how it aligns:

Stack Layer	Role in Workplace Enablement
Layer 3: User Experience	Ensures all touchpoints (apps, signage, check-in) are intuitive and inclusive.
Layer 4: Applications & Services	Booking engines, user dashboards, and service layers interface with back-end systems.
Layer 5: Data & Context	Tracks occupancy, usage trends, user behaviours, and context for decision-making.
Layer 6: Integration Layer	Connects desk sensors, calendars, wayfinding maps, and building systems seamlessly.
Layer 7–8: Systems & Physical Environment	Provides the foundational infrastructure — sensors, screens, power, and connectivity.

By addressing each layer, organisations avoid fragmented deployments and instead deliver a unified, user-centric environment that boosts productivity, enhances flexibility, and supports long-term space planning.

Hybrid Work and Seamless Digital-Physical Transitions

The shift toward hybrid working has redefined the purpose and function of the workplace. No longer a default daily destination, the office is now a deliberate choice — a place people visit for connection, collaboration, or focused work that can't be replicated at home. To stay relevant and valued, workplaces must offer added experiential value, removing friction from the workday and ensuring continuity between digital and physical environments.

This requires a fundamental rethinking of how space, technology, and human behaviour intersect. It's no longer enough to simply have video conferencing capabilities or a hot desk policy. Truly smart hybrid workplaces are responsive ecosystems, shaped by data, integrated systems, and a human-centric design philosophy. They make transitions feel effortless — from remote to in-person, solo to team-based work, or digital to physical.

At the heart of this transformation lies the 360 Smarter Stack methodology, which provides a structured framework to deliver a workplace capable of supporting seamless transitions, underpinned by both strategic intent and robust technical infrastructure.

Bridging the Digital and Physical Experience

Hybrid employees frequently shift contexts — between locations, devices, and types of tasks. As a result, the workplace must function as part of a broader experience ecosystem.

Employees expect:

- The same level of digital convenience in the office that they have at home.

- Continuity of data and tools across all environments.

- Access to services like space booking, wayfinding, or climate control via personal devices.

This requires workplaces to be digitally enabled in every sense — integrating IoT sensors, workplace apps, room systems, and analytics platforms into a cohesive environment.

Examples include:

- Smart room scheduling that adapts to changing meeting times or attendee locations.

- Touchless access and check-in systems that reduce wait times and enhance security.

- Personalised environmental controls (like lighting and temperature) that respond to individual preferences.

The goal is to create a workplace that anticipates needs, removes logistical barriers, and feels as intuitive as working from home.

Smart, Adaptive Office Design

Hybrid-ready spaces must support a wide spectrum of work styles. This includes not only collaborative settings but also private, focused, and transitional zones. Some key spatial elements include:

- Reservable Focus Pods: Private, soundproof booths with integrated AV and booking sensors.

- Agile Collaboration Spaces: Reconfigurable rooms with mobile furniture and whiteboards, supported by intelligent AV that accommodates remote participants equally.

- Quiet Zones: Designed for heads-down work with minimal distractions.

- Community Hubs: Social spaces that support informal networking and team culture.

These spaces become even more powerful when data from sensors and usage analytics are used to fine-tune their availability, capacity, and performance.

Mobile-First Interfaces for a Connected Experience

In the hybrid workplace, smartphone apps and digital interfaces become the remote control for everything from desk bookings to climate preferences. A well-integrated workplace app should:

- Offer a dashboard of the day, highlighting booked spaces, visitors, events, and suggested nearby teammates.

- Provide real-time wayfinding and alerts on room availability or congestion.

- Enable service requests (IT help, catering, maintenance) with integrated tracking.

- Allow employees to log feedback, helping FM and IT teams respond dynamically.

These tools should work in unison with the physical environment, enabling personalisation at scale, driven by identity and context.

Real-Time Data Sync and Environmental Responsiveness

A seamless hybrid experience hinges on real-time data flow across systems. This includes:

- Calendar integrations to auto-book resources or sync collaboration tools.

- Workplace intelligence that adjusts lighting, temperature, or AV settings based on occupancy and preferences.

- Sensor networks that detect patterns and predict demand — for instance, increasing ventilation in crowded areas or redirecting employees to quieter zones.

By orchestrating these systems through a common integration layer — a key component of the 360 Smarter Stack — organisations can create fluid, intelligent spaces that respond to changing user behaviour instantly.

Culture, Community, and Meaningful Presence

Even with all the right tools and systems, the smart workplace must also deliver emotional and cultural value. Infrequent office visits mean every interaction matters more. The office becomes a place to reinforce values, inspire creativity, and foster community.

This means:

- Designing experiences that reward in-person collaboration (e.g., immersive brainstorm spaces or wellness amenities).

- Showcasing commitment to ESG goals through smart energy use, biophilic design, and sustainable materials.

- Using digital signage and displays to promote team wins, DEI initiatives, and employee stories, reinforcing a sense of belonging.

Smart hybrid workplaces are not just about efficiency — they're about meaningful presence. When technology, space, and people are synchronised, the workplace becomes a magnet, not a mandate.

Layer 4: Applications & Services

Chapter 8: Selecting Smart Office Applications

The smart office experience comes to life through the applications people interact with every day — from reserving a workspace or starting a video meeting, to monitoring energy use or accessing secure areas. These tools represent the **visible tip of the iceberg** within the 360 Smarter Stack — the layer where users experience the benefits of everything that lies beneath: data, integration, infrastructure, and vision.

Choosing the right suite of applications is a critical decision. It impacts user satisfaction, operational efficiency, energy consumption, and the overall ROI of a smart transformation. But with a constantly evolving landscape of platforms, apps, and tools, it's easy to be overwhelmed by choice or misled by marketing hype.

This chapter explores the real-world capabilities to look for, the role of integration and data flow, and how to make smart application choices that align with your long-term goals — rather than creating short-term fixes or future silos.

Space Booking, AV Control, Energy Dashboards, Security Tools

Space Booking Systems: Empowering Flexibility with Data Intelligence

Modern workspaces are fluid. Staff might be in the office on Mondays and Thursdays, contractors may need temporary space, and leadership may require access to executive rooms at short notice. Smart space booking platforms are critical for orchestrating all of this efficiently.

But beyond merely reserving a seat or room, these systems have evolved into intelligent orchestration tools. They form a central part of the user experience layer while pulling data and functionality from multiple deeper stack layers.

Key capabilities include:

- Interactive floor plans with real-time space status, visible via web or mobile apps.

- Auto-cancellation of no-show bookings using occupancy sensors, increasing utilisation.

- Zone-based or team-based booking logic, supporting flexible neighbourhood models.

- Visitor integration, allowing external guests to be linked to host bookings and check-ins.

- Environmental data overlays — showing rooms with preferred temperature, noise levels, or lighting setups.

- Predictive analytics — helping organisations right-size their space portfolio, redesign underutilised areas, or forecast future needs.

From an operational perspective, these tools feed valuable data into FM and real estate strategies, influencing everything from cleaning rotas to capital expenditure planning.

AV Control Systems: Frictionless Collaboration Across Locations

As hybrid meetings become the norm, the expectation for seamless, frustration-free AV has never been higher. Users want to walk into a room and be up and running in seconds — with no cable hunting, no manual switching, and no IT rescue calls.

Smart AV control platforms are the nerve centre of this experience. But unlike legacy setups where each component operated independently, modern AV systems are part of a connected ecosystem that:

- Automatically recognises room bookings and powers up displays and cameras.

- Launches the correct video conferencing platform (Teams, Zoom, etc.) based on the invite.

- Provides unified control via in-room tablets, touch panels, or mobile apps.

- Adjusts room settings based on user profiles or meeting types — for example, lowering blinds and dimming lights for a presentation.

- Offers remote monitoring, diagnostics, and reboot options via central management tools, reducing downtime and IT tickets.

- Captures usage data and performance metrics to guide future upgrades or support decisions.

This integration isn't just technical — it's cultural. A well-designed AV control system eliminates anxiety and increases confidence in hybrid meetings, fostering inclusivity for remote and in-person attendees alike.

Energy Dashboards: Visualising and Acting on Environmental Impact

As ESG (Environmental, Social, and Governance) strategies move from aspiration to execution, energy dashboards become a crucial component of smart workplace accountability.

At their core, these dashboards provide real-time and historical visibility into energy consumption across your building or campus. But the real value lies in how they combine and contextualise data:

- Breakdowns by floor, zone, system, or even individual device help FM teams target interventions precisely.
- Overlaying occupancy data (from booking systems or motion sensors) reveals mismatches between usage and consumption.

- Dynamic alerts highlight anomalies — such as air conditioning running in an unoccupied room.

- User-facing displays on digital signage reinforce sustainable behaviours and communicate progress to staff and visitors.

- Integration with building management systems (BMS) enables automated optimisation — for example, reducing lighting or HVAC in low-traffic areas.

More advanced dashboards also incorporate carbon footprint tracking, helping organisations move toward Net Zero targets and submit robust ESG reporting. This isn't just a facilities tool — it's a strategic asset.

Security and Access Tools: Enabling Trust and Autonomy

Security has long been viewed as a gatekeeper — something that protects, restricts, and audits. But in a smart workplace, it must also enable freedom of movement, autonomy, and user empowerment, while still maintaining a secure environment.

Modern smart security systems are data-driven, flexible, and seamlessly integrated into the broader ecosystem. Key capabilities include:

- Mobile-based access using NFC, Bluetooth, or biometric credentials — eliminating the need for physical cards.

- Role-based access policies dynamically adjusted based on department, seniority, or temporary visitor status.

- Real-time integration with booking systems — for example, only granting meeting room access when a booking is active.

- Event-based camera activation and analytics-powered alerts in critical areas or after hours.

- Integration with HR systems and identity management platforms, ensuring personnel changes are reflected in access privileges immediately.

- Cross-system visibility — e.g., connecting access logs with occupancy data to detect suspicious patterns or improve emergency response.

From lone worker safety to pandemic response readiness, modern access control is no longer just about locks and readers. It's about building a secure, responsive, and intuitive workplace experience.

Evaluating platforms and ecosystems (vs point solutions)

Evaluating Platforms and Ecosystems (vs Point Solutions)

In the evolving smart office landscape, the tools you choose to power your workplace are as critical as the infrastructure you build them on. One of the most strategic decisions in the journey toward a smart environment is whether to adopt an integrated ecosystem of platforms or to assemble best-in-class point solutions. This choice will fundamentally shape the user experience, operational model, support structure, and long-term viability of your smart building program.

The Platform Ecosystem Approach

A platform or ecosystem refers to a comprehensive, modular environment where multiple functions (e.g. booking, access control, lighting management, energy usage, digital signage, etc.) are built or integrated into a unified system — usually with a common user interface, data structure, and support model.

These platforms are often designed to cover a broad spectrum of workplace needs, enabling a more connected, scalable, and manageable environment. Some leading platforms include:

- **IWMS (Integrated Workplace Management System) Platforms**
- **Smart Workplace Experience Suites**
- **Building Operating Systems**
- **Enterprise Integration Platforms**

1.IWMS (Integrated Workplace Management System) Platforms

An IWMS platform is a comprehensive software suite that helps organisations manage and optimise their workplace resources, real estate, and facilities. These platforms are used to streamline operations across the full lifecycle of buildings and office environments — including space management, maintenance, energy, sustainability, and capital planning.

Key Functions of IWMS Platforms:

- Space Management – Track and plan how office space is used, including desk booking, space utilization, and reconfiguration.

- Real Estate Portfolio Management – Monitor leases, occupancy costs, and property performance.

- Maintenance Management – Schedule and track preventive and reactive maintenance across assets and systems.

- Project Management – Manage renovations, fit-outs, or relocation projects.
- Sustainability & Energy Management – Monitor energy usage, emissions, and environmental performance (e.g., for ESG reporting).

- Workplace Services – Enable services such as room reservations, service requests, and visitor management.

Examples of IWMS Platforms:
- Archibus (by Eptura)
- Planon
- FM:Systems
- Trimble Manhattan
- iOFFICE
- MCS Solutions (Spacewell)

These platforms are particularly common in large enterprises, government bodies, universities, and real estate-heavy industries, where managing cost, compliance, and efficiency across many sites is a high priority.

2.Smart Workplace Experience Suites

These suites act as a single interface—often a mobile or web app—that brings together multiple tools and services used across the workplace. They're built to make the digital-physical workplace seamless, intuitive, and responsive.

Core Capabilities

Smart workplace experience suites typically include:

1 Desk & Room Booking

Reserve spaces, desks, or amenities in real time.
See who else is in the office and where they're sitting.

2   Wayfinding & Navigation
Interactive maps that guide users to meeting rooms, colleagues, or points of interest.

3   Mobile Workplace Apps
One-tap access to services like booking, support requests, feedback, and company news.

4   Hybrid Work Scheduling
Plan and coordinate office attendance with your team.

5   Service Requests & Feedback
Raise IT/FM support tickets, report issues, or give feedback on services.

6   Integration with Building Systems
Connect to HVAC, lighting, and access control for personalized, sustainable environments.

7   Analytics & Insights
Capture space usage, satisfaction metrics, and behavioural trends to inform workplace strategy.

How They Differ from IWMS Platforms

Feature	IWMS	Smart Workplace Experience Suite
Focus	Real estate, assets, FM ops	User experience, engagement
Primary Users	FM, CRE, IT teams	Employees, visitors, managers
Interface	Often admin-focused	Mobile-first, intuitive UX
Integration	Back-end systems, lifecycle	Front-end touchpoints, UX design

Examples of Smart Workplace Experience Suites

- HqO
- MODO Labs
- JLL Jet
- Equiem
- Appspace
- Envoy
- Proxyclick (for visitor management)

Why They Matter for Smart Office Strategy

In the 360 Smarter Stack, smart workplace experience suites sit at the User Experience and Applications & Services layers. They are critical for:

1. Adoption of smart office tools

2. Personalised, human-centric workplaces

3. Driving ROI through improved satisfaction and engagement

They don't just connect systems—they connect people to place.

3.Building Operating Systems

A Building Operating System is a middleware platform that connects disparate building systems—like HVAC, lighting, access control, energy monitoring, and IoT sensors—into a single, interoperable framework.

Think of it as the "Windows" or "iOS" of a building: it doesn't replace your hardware or applications, but it allows them to work together more cohesively.

Core Functions of a BOS

- Device & System Integration
  - Connects legacy systems (e.g., BMS/BAS) with modern IoT devices using open protocols like BACnet, Modbus, MQTT, and API connectors.

- Real-Time Data Aggregation
  - Collects data from all connected systems and normalises it into a common data model.

- Analytics & Insights
  - Delivers actionable dashboards for energy efficiency, occupancy trends, and asset performance.

- Control & Automation
  - Enables advanced control logic (e.g., room temperature based on occupancy, lighting based on daylight levels).

- User & Tenant Engagement
  - Some platforms (e.g., Spica) include front-end experiences for tenants, such as mobile apps or workplace experience layers.

- Security & Access Governance
  - Centralises identity and access policies, allowing for unified management across multiple systems.

Examples of Leading BOS Platforms

Spica Technologies: Offers smart workplace tools alongside building control features. Strong in occupancy analytics, ESG reporting, and workplace experience.

BuildingOS (by Acuity Brands): Focused on energy management and sustainability reporting, ideal for ESG-aligned portfolios.

SmartSpace: Combines space utilisation, IoT integrations, and workplace experience features.

Position in the 360 Smarter Stack

Stack Layer	BOS Role
Data & Context	Aggregates and models data from building systems
Integration Layer	Acts as middleware/API hub to connect systems
Applications & Services	Powers automation, analytics, and sometimes user apps
Systems & Devices	Interfaces directly with building hardware (HVAC, lighting, etc.)

Why BOS Platforms Matter

- They unlock true interoperability in modern smart buildings.
- Reduce vendor lock-in by enabling open, flexible architectures.

- Act as a foundation for scalable smart strategies, from pilot to portfolio level.

- Help bridge the gap between technical complexity and operational value.

4.Enterprise Integration Platforms

These platforms provide a low-code or no-code environment to automate workflows, unify data, and create business apps that span across both smart building infrastructure and traditional enterprise systems (e.g., HR, finance, CRM).

Unlike a typical Building Operating System that focuses on infrastructure and environmental systems, Enterprise Integration Platforms extend smart building capabilities into the fabric of business operations.

Core Capabilities

- Data Unification & Transformation
    - Connect to multiple systems (e.g., IoT sensors, ERP, CRM, Active Directory) and bring data into a central, structured format.
    - Use semantic modelling and logic layers to make data meaningful and actionable.

- Workflow Automation
    - Trigger automated actions based on smart building events:
    - If room occupancy exceeds a threshold → notify facilities team.
    - If energy usage spikes → create a service ticket or alert.

- Custom App Development
    - Build mobile and desktop apps that interact with smart office data:
    - Room booking tools
    - Maintenance dashboards
    - Visitor check-in flows

- Visualisation & Reporting
    - Use tools like Power BI (within Microsoft Power Platform) to visualise key KPIs—space utilisation, ESG metrics, energy savings.

- Security & Governance
    - Enterprise-grade controls, role-based access, and secure integrations with identity platforms like Azure AD.

Platform Examples:

Microsoft Power Platform

- Includes Power BI, Power Automate, Power Apps, and Dataverse.
- Ideal for connecting smart building systems with Microsoft 365 tools (Teams, Outlook, SharePoint).
- Common use cases: automatic notifications, custom workplace apps, ESG dashboards.

Cisco Spaces

- Built specifically to harness the intelligence of Cisco's networking infrastructure.
- Provides real-time location services, occupancy insights, and API integration for third-party applications.
- Excellent for hybrid work management, hot desking, and occupancy-driven services.

Position in the 360 Smarter Stack

Stack Layer	Enterprise Integration Role
Integration Layer	Acts as the glue between smart systems and enterprise platforms
Data & Context	Helps unify, enrich, and present data meaningfully
Applications & Services	Powers apps and automations that end-users and ops teams interact with
Operations & Culture	Supports governance, compliance, and change management frameworks

Why These Platforms Matter

- Accelerate Time-to-Value: Enable rapid prototyping and deployment of smart solutions without deep technical skills.

- Enable Business Agility: Let organizations adapt quickly to workplace changes or strategic goals.

- Support Scalable Integration: Easily connect dozens of systems—from IoT to finance to HR—without building from scratch.

- Drive ROI: By linking smart building data to business performance indicators, these platforms help prove (and improve) ROI.

Key benefits adopting the Platform Ecosystem Approach include:

- Consistent user experience: Staff and visitors interact with one intuitive interface, reducing training, confusion, and friction.

- Centralised data insights: All systems feed into a common data layer, enabling unified dashboards, alerts, and reporting — ideal for ESG tracking, occupancy planning, or predictive maintenance.

- Stronger cybersecurity and governance: Platforms are often more mature in terms of patching, permissions management, data privacy, and audit trails.

- Easier upgrades and scalability: Adding new features, buildings, or user groups is more straightforward within an ecosystem.

- Vendor accountability: A single supplier or consortium can take end-to-end responsibility for delivery and support.

These benefits align strongly with the 360 Smarter Stack approach, particularly across the Integration Layer, Data & Context, and Applications & Services layers.

Point Solutions: Precision and Agility

Point solutions, in contrast, focus on solving a specific problem exceptionally well — such as the best-in-class room booking tool, a cutting-edge occupancy sensor, or an advanced AV scheduling platform. These tools are often preferred in organisations that need:

- Niche functionality that generalist platforms can't offer.

- Speed to value — rapid deployments to meet urgent business needs.

- Trial or proof-of-concept scenarios before committing to larger investments.

- Modular investment strategy — rolling out capabilities incrementally.

Examples might include:

- Desk booking – e.g. Condeco, Robin, Skedda

- Visitor management – e.g. Proxyclick, SwipedOn

- AV control – e.g. Crestron Fusion, Q-SYS

- Environmental sensors – e.g. VergeSense, Disruptive Technologies

While powerful, point solutions come with clear trade-offs:

- Data silos: Each tool may store data in its own format, requiring custom integrations for reporting and analysis.

- User friction: Staff may have to log into multiple apps or interfaces, often duplicating tasks (e.g., booking a room and separately booking catering or AV).

- Integration overhead: APIs, middleware, and interface development become essential, increasing both upfront costs and ongoing support needs.

- Scalability issues: Point solutions that work well in one building may become difficult to scale across a global portfolio.

- Fragmented governance: Security policies, usage standards, and support contracts may vary widely between tools.

Strategic Considerations: Finding the Right Fit

In practice, the right approach often lies somewhere between the two extremes. Most successful smart workplace programs adopt a hybrid model, where:

- A core platform acts as the backbone for booking, identity management, and analytics.

- Selective point solutions are integrated where specialist capabilities are required.

- Integration and governance layers ensure coherence, security, and data continuity across the stack.

To guide this decision, consider the following questions:

Evaluation Criteria	Key Questions
Vision Alignment	Does the platform support our long-term workplace strategy and ESG goals?
User Experience	Will users benefit from a single interface, or is their role-specific tool more effective?
Data Integration	Can all tools share data effectively, or will silos emerge?
Security & Compliance	Are you comfortable managing multiple vendors and their individual risks?

Support Structure	Do you have internal skills and bandwidth to manage integration, or is simplicity preferred?
Vendor Viability	Will the vendor grow with your organisation or lock you into a narrow path?

Alignment with the 360 Smarter Stack

The 360 Smarter Stack provides a structured lens through which to evaluate platforms and tools. It encourages leaders to assess technologies not in isolation, but in terms of how they:

- Integrate across layers (e.g. from Devices to Data to Applications)

- Contribute to User Experience and Operational Support

- Enable strategic outcomes around ESG, productivity, and engagement

A smart platform that fails to support a layered, connected delivery model may appear attractive initially but can create fragmentation over time. Conversely, a well-chosen ecosystem can act as the digital nervous system of the building — responsive, coordinated, and insightful.

Chapter 9: Orchestrating Services

As the complexity of smart offices grows, so does the need for intelligent coordination between systems. The real power of a smart workplace doesn't come from isolated technologies—it emerges when multiple systems work together in harmony, delivering seamless and automated experiences for users and operators alike.

Service orchestration is the connective tissue between apps, devices, and enterprise platforms. It enables a single user action—like booking a meeting room—to trigger a cascade of automated tasks across facilities, IT, and business layers. This chapter explores how workflows are designed, integrated, and managed, and how a future-ready architecture supports long-term scalability.

Use Cases and Workflows Across Apps

(e.g., booking a room triggers HVAC/lighting)

In a truly intelligent workplace, the real magic doesn't come from deploying isolated technologies—it emerges when systems speak to one another, orchestrating coordinated actions that simplify tasks, enhance comfort, and drive business value. This is where smart workflows transform everyday moments into seamless experiences.

Smart offices don't just have digital tools—they connect them through carefully designed trigger-based automations, underpinned by sensor data, real-time context, and business rules. Below, we explore a range of high-impact use cases that illustrate this orchestration in action—each tied to different layers of the 360 Smarter Stack, and each demonstrating how smart technology can evolve from being functional to being transformational.

1. Smart Meeting Room Booking and Automation

Scenario: An employee books a meeting room for a hybrid client presentation.

Orchestration Workflow:

- The employee uses a centralised booking platform (web or app) to reserve the room, selecting AV needs, number of attendees, and whether guests will join remotely or in person.

- The calendar integration (e.g. Outlook or Google Workspace) sends out invites and connects the meeting to in-room collaboration tools (e.g. Teams, Zoom).

- Five minutes before the meeting, AV systems automatically power on, adjust to the selected configuration, and preload presentation settings.

- Lighting and climate control adjust to a comfortable pre-set for occupancy.

- If visitors are attending, security access is pre-authorised, and a QR code is sent to their phone via email.

- Upon arrival, digital signage shows the room status, while wayfinding kiosks guide guests to the correct floor or room.

- If the room is not occupied within 10–15 minutes, sensors flag it as a "no-show" and the room is automatically released back into availability.

- Facilities and workplace teams receive usage analytics, helping them plan for future optimisation.

This is more than just convenience—it reduces wasted space, supports ESG goals through energy efficiency, and improves the visitor experience with minimal human intervention.

2. Predictive Maintenance and Automated Facilities Response

Scenario: A lift motor shows abnormal vibration levels during normal operation.

Orchestration Workflow:

- IoT vibration and temperature sensors detect early-stage wear on the motor.

- The building management system (BMS) triggers an alert, forwarding telemetry data to the CAFM/IWMS platform (e.g. Planon, IBM TRIRIGA).

- The system auto-generates a maintenance ticket with diagnostics and asset history.

- A mobile alert is sent to the on-site engineer, including recommended parts and step-by-step instructions.

- The platform checks spare part inventory and flags whether additional components need ordering.

- A dashboard updates in real time to show issue resolution progress and any downtime impact.

- If the system is linked with an energy dashboard, power draw is minimised automatically to avoid further strain.

This level of orchestration reduces downtime, extends asset life, and moves facilities management from reactive to proactive, aligning with lean operational practices.

3. Seamless Visitor Journey and Hospitality Automation

Scenario: A senior stakeholder from a partner organisation is scheduled to visit the office.

Orchestration Workflow:

- The employee hosting the visitor logs them into a visitor management system and selects meeting location and time.

- The visitor receives a branded welcome email with a secure QR code for check-in, Wi-Fi access, and car park guidance.

- Upon arrival, the guest scans their code at a digital kiosk, triggering a notification to the host via Teams or SMS.

- Access control systems grant limited access to lifts and rooms based on time and location.

- Pre-ordered refreshments are automatically notified to catering, while any AV requirements are pre-loaded into the meeting room.

- Post-visit, a feedback survey is automatically emailed, and their temporary access is revoked.

This automated journey feels premium and professional for visitors while ensuring security and efficiency for the host organisation.

4. Hybrid Working and Agile Workspace Enablement

Scenario: An employee is working from the office two days per week and needs to book a desk and collaborate with colleagues.

Orchestration Workflow:

- Via a mobile app, the employee books a desk in a "quiet zone" and a collaborative huddle space for the afternoon.

- Booking data is analysed across floors, and only occupied zones are powered up for HVAC and lighting.

- The desk's occupancy sensor confirms arrival and unlocks monitor and power settings.

- If the space has smart lockers, a personal locker is assigned based on proximity.

- Real-time data feeds back into the IWMS to update space availability dashboards.

- Insights from this behaviour inform long-term workplace planning, cleaning schedules, and hybrid attendance patterns.

These small daily actions, when connected through orchestration, power smart space planning and make hybrid working frictionless.

5. Access Control and Just-In-Time Security Provisioning

Scenario: A contractor needs access to the building for scheduled maintenance.

Orchestration Workflow:

- A request is submitted via an integrated contractor management system, with ID and credentials checked in advance.

- Once approved, temporary access credentials are issued for specific zones, times, and days—delivered to the contractor's phone.

- On arrival, they check in using a QR or face ID system, which automatically updates the building access logs.

- If their presence overlaps with an unauthorised time or area, alerts are triggered in the security platform.

- When the maintenance window closes, access rights are automatically revoked, and all interactions are recorded for audit.

This workflow ensures security compliance, reduces administrative overhead, and supports zero-trust access strategies.

6. Energy & Sustainability Optimisation

Scenario: The business is aiming to cut energy costs and improve environmental performance.

Orchestration Workflow:

- Real-time sensor data from lighting, HVAC, plug loads, and occupancy flows into a central energy dashboard.

- During low-occupancy periods or evenings, the system identifies underused zones and powers them down autonomously.

- Based on weather forecasts and booking data, the system dynamically adjusts heating/cooling profiles to maximise efficiency.

- Employees receive prompts encouraging them to book desks in high-efficiency areas.

- ESG teams can access dashboards to track carbon savings, emissions, and compliance KPIs.

These orchestrations contribute meaningfully toward corporate sustainability goals, with minimal impact on employee comfort or operations.

7. Emergency Response and Life Safety

Scenario: A fire alarm is triggered on Level 5 during working hours.

Orchestration Workflow:

- The fire detection system sends an instant notification to all building systems.

- Access control points are unlocked automatically to allow safe egress.

- Digital signage and wayfinding displays update with animated exit guidance.

- Occupancy sensors help verify who is still inside the building, alerting emergency responders if anyone is in a high-risk area.

- Emergency lighting activates in affected zones, while lifts are grounded for safety.

- All activity is logged in the building's incident response system for post-event analysis.

In such moments, orchestrated systems can save lives—by delivering clarity, speed, and control across the building ecosystem.

Why It Matters

These use cases show that smart workplace orchestration is not about one app doing one thing well—it's about the choreography of connected systems, working together in real time, guided by data, context, and purpose.

Benefits of orchestrated workflows include:

- Reduced friction in daily operations.

- Better space, energy, and asset utilisation.

- Safer and more secure environments.

- Higher-quality visitor and employee experiences.

- Improved resilience and business continuity.

By defining and designing these workflows early in a smart office transformation, organisations ensure that technology investments are not just modern but meaningfully aligned with both human needs and strategic outcomes.

Future-ready architecture

As organisations increasingly invest in smart building technologies, the importance of designing for scalability, adaptability, and longevity cannot be overstated. A smart office is not a static endpoint—it is a platform for continuous improvement. This requires a future-ready architecture that can flex with business change, accommodate emerging technologies, and avoid being trapped by vendor lock-in or short-term tactical decisions.

Smart workplaces must be designed not just for today's needs, but with the foresight to evolve over time. That means taking a layered, modular, and standards-based approach—ensuring that the systems deployed are interoperable, secure, and capable of integrating across future platforms, devices, and user expectations.

1. Modular & Layered Technology Stacks

A future-ready architecture starts with a well-defined technology stack, divided into functional layers:

- Device Layer: IoT sensors, occupancy counters, smart meters, beacons, and gateways.

- Edge Layer: Local controllers and compute (e.g. lighting control hubs, HVAC edge processors).

- Connectivity Layer: Wired and wireless networks, including secure segmentation for OT and IT systems.

- Platform Layer: IWMS, BMS, space management, AV control, and analytics platforms.

- Integration Layer: APIs, middleware, and enterprise integration tools (e.g. Microsoft Power Platform, Mulesoft, Cisco Spaces).

- Experience Layer: End-user apps, kiosks, digital signage, and portals for employee interaction.

By architecting across these layers—like the 360 Smarter Stack methodology—organisations gain the ability to swap, upgrade, or extend individual layers without reworking the entire environment. This separation of concerns allows for agile scaling and targeted innovation.

2. API-First and Standards-Based Integration

Many early smart building initiatives were hindered by proprietary systems, which offered limited connectivity and were resistant to change. A future-ready architecture prioritises:

- Open APIs for integration across platforms.

- Support for industry standards and protocols such as BACnet, MQTT, OPC-UA, and KNX.

- Use of data models that align with frameworks like RealEstateCore, Haystack, or Brick Schema.

These standards provide a foundation for consistent data exchange, simplify onboarding of new tools, and enable more powerful orchestration across systems (e.g., tying access control to energy dashboards or occupancy trends).

3. Cloud-Native & Hybrid Deployment Models

Modern architecture should consider hybrid cloud models, allowing organisations to deploy systems on-premises, in the cloud, or across both—based on business needs, data sovereignty, and latency.

Benefits of cloud-native or hybrid architecture include:

- Elastic scalability as device counts grow.

- Simplified remote management of multi-site estates.

- Access to AI/ML-powered insights via cloud-based analytics engines.

- Faster deployment of feature updates and patches.

- Reduced reliance on local servers and infrastructure.

Edge computing still plays a role—particularly for real-time systems like BMS or AV—but the long-term trend is toward cloud-hosted platforms with intelligent edge nodes.

4. Security and Privacy by Design

Security is a critical pillar of any future-ready smart office. With the convergence of IT and OT, and the proliferation of connected devices, the attack surface expands. Architecture must embed cybersecurity and data privacy from the outset, including:

- Zero trust access control models across all devices and systems.

- Regular vulnerability assessments and patching regimes.

- Encryption at rest and in transit for all data.
- Segregation of building systems from corporate networks.

- Compliance with regulations like GDPR, especially when processing occupancy or biometric data.

A secure architecture protects both the organisation and the people who use the space.

5. Extensibility and Integration with Enterprise Ecosystems

Future-ready smart offices don't operate in isolation—they integrate with wider enterprise ecosystems, such as:

- HR systems (for onboarding/offboarding workflows).

- Facilities and FM software (e.g. Planon, Archibus).

- IT service management platforms (e.g. ServiceNow).

- ESG reporting tools (e.g. Envizi, Measurabl).

- Collaboration suites (e.g. Microsoft 365, Google Workspace).

This extensibility ensures smart systems contribute to business-wide value—linking space utilisation to HR policy, or energy performance to ESG commitments.

6. AI and Machine Learning Readiness

Future-ready infrastructure should be AI-enabled by design. That means:

- Capturing rich data sets from sensors, usage logs, and user behaviour.

- Using cloud-based analytics engines (e.g. Azure Synapse, AWS IoT Analytics).

- Creating feedback loops that allow systems to learn and adapt—optimising energy use, predicting maintenance needs, or personalising experiences.

Machine learning thrives in well-architected environments. A smart office becomes smarter over time when designed to ingest and act on data.

7. Digital Twins and Simulation Tools

An emerging frontier in future-ready architecture is the use of digital twins—virtual models of physical spaces and systems that allow simulation, analysis, and optimisation in real-time.

Digital twins can:

- Simulate occupancy flows and identify bottlenecks.
- Test HVAC strategies against weather forecasts.
- Predict the impact of changes to layout, policy, or equipment.

When integrated with the rest of the stack, digital twins provide a powerful tool for scenario planning, sustainability, and continuous improvement.

A future-ready architecture is not just a technical ambition—it is a strategic enabler. It provides the agility to meet tomorrow's demands, the resilience to adapt to change, and the foundation to continually enrich the workplace experience.

By adopting layered design principles, leveraging open standards, and aligning with the 360 Smarter Stack, organisations can de-risk their investment and build a workplace platform that grows with them—technologically, operationally, and culturally.

Future-Ready Architecture

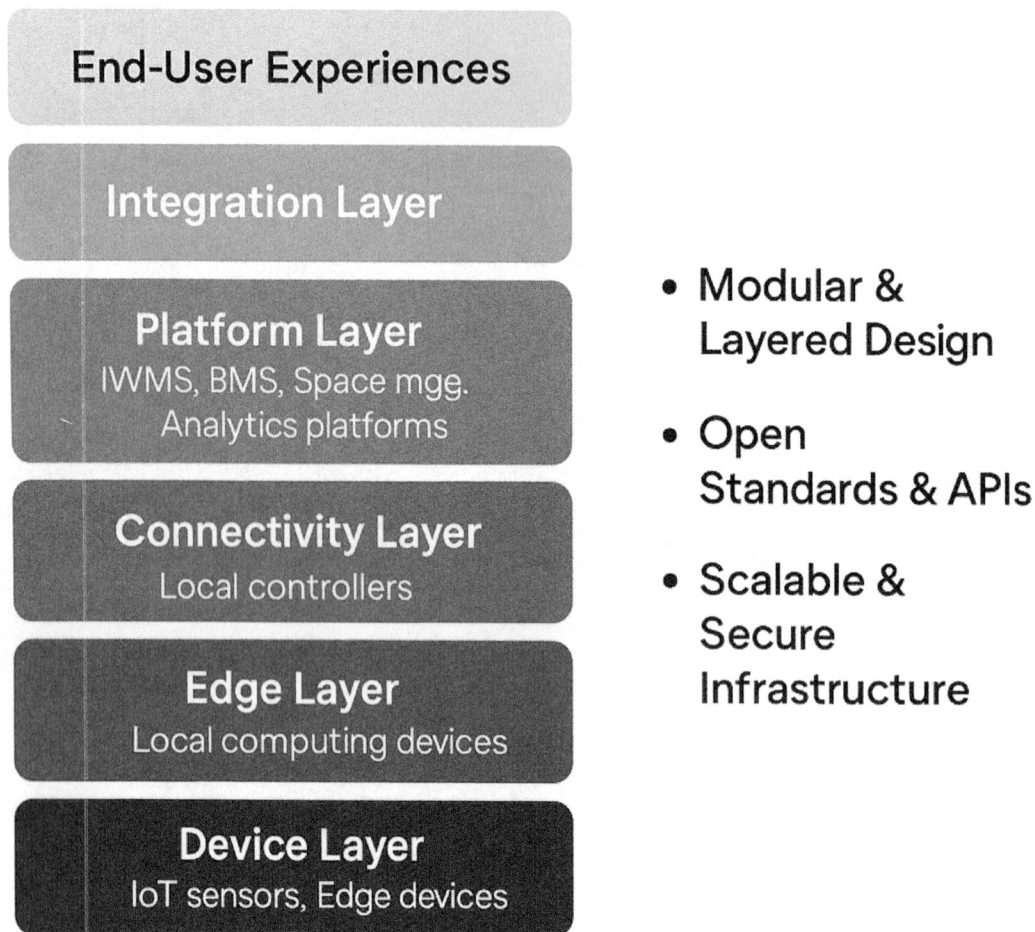

End-User Experiences

Integration Layer

Platform Layer
IWMS, BMS, Space mge.
Analytics platforms

Connectivity Layer
Local controllers

Edge Layer
Local computing devices

Device Layer
IoT sensors, Edge devices

- Modular & Layered Design

- Open Standards & APIs

- Scalable & Secure Infrastructure

Diagram: Modular & Layered Technology Stacks

Layer 5: Data & Context

Chapter 10: Building a Contextual Data Model

The smart office is only as intelligent as the data it understands — and to extract meaningful insights, raw data must be structured, contextualised, and made interoperable. Building a contextual data model is about moving from disconnected data points to a dynamic ecosystem of information that fuels automation, personalisation, and strategic decision-making. It ensures that every device, space, and interaction within a smart building contributes to a coherent, data-driven picture.

At the heart of this transformation is a focus on **data tagging**, **semantic layers**, and **system interoperability** — which together form the backbone of a usable and scalable smart infrastructure. In this chapter, we explore how this model enables actionable insights and empowers organisations to make informed, timely, and automated decisions across the workplace.

Data Tagging, Semantic Layers, and Interoperability

The ability to harness data effectively is one of the most important differentiators in delivering a truly smart office. However, the mere presence of data is not enough. In most buildings, data is plentiful — generated by thousands of sensors, devices, and systems — yet it remains underutilised, disjointed, and hard to act upon. The key to unlocking its full potential lies in contextualisation — understanding what the data represents, where it comes from, and how it can be used in broader business logic.

This contextualisation relies on three foundational pillars: **data tagging**, **semantic layering**, and **interoperability**. These are not abstract technical concepts; they are practical enablers of smart services, automation, and meaningful insights — and they form the invisible scaffolding of any intelligent workplace infrastructure.

Data Tagging – Structuring the Fundamentals

Think of data tagging as adding labels to your digital world — a way of saying "this is what this data is, where it is, and what it's for." Every connected device, system, and data stream in a smart building should have clear, consistent metadata that defines its characteristics and context.

For example:

- A CO_2 sensor on the third floor in a meeting room might be tagged with:
    1. Building: HQ
    2. Level: 3
    3. Zone: West
    4. Space Type: Meeting Room
    5. Device Type: CO_2 Sensor
    6. Asset ID: 03321
    7. Function: Air Quality

By standardising this kind of metadata, the system can automatically organise, search, and query data sources. Tagging enables:

- Faster onboarding of new assets or spaces

- Easier maintenance and fault diagnostics

- Enhanced automation, where the system can dynamically adjust based on what and where something is

Tagging can follow open standards like Project Haystack, which provides a taxonomy and structure for building-related data points. These kinds of standards also make it easier to integrate with other platforms in the future, ensuring long-term flexibility and scalability.

Semantic Layers – Creating Meaningful Relationships

While tagging tells us what something is, semantic modelling tells us how things relate. A semantic layer is a logic model that defines how data elements interact and belong within a building's digital twin. It allows the system to reason, infer, and behave intelligently.

For instance:

- The system might understand that Room 201 is part of the "Client Zone" on Floor 2, and that it's a "Formal Meeting Room" associated with Business Unit X.

- It knows that this room includes five data points: occupancy sensor, lighting circuit, temperature control, booking status, and user satisfaction score — and that they all contribute to a "comfort index" for that space.

Semantic modelling makes these relationships explicit and machine-readable, so systems can respond in ways that align with business logic, rather than isolated data triggers. This enables use cases such as:

- Turning off lighting when the room is not booked and unoccupied

- Correlating booking frequency with air quality levels

- Personalising room settings based on expected user profiles

Semantic models are often built using ontologies like Brick Schema, RealEstateCore, or IFC (Industry Foundation Classes), which describe building spaces, systems, and equipment in a structured way. These ontologies promote shared understanding across platforms and stakeholders, making collaboration easier and reducing duplication of effort.

Interoperability – The Backbone of Integration

Interoperability is what enables all this data to be shared, processed, and utilised by various systems across your smart stack. Without it, even the best-tagged and semantically modelled data sits in isolated silos, disconnected from broader workflows and insights.

Achieving interoperability requires:

- Common data formats and exchange protocols, such as BACnet, KNX, MQTT, OPC UA, and RESTful APIs.

- Middleware or integration platforms, which act as translation layers between disparate systems (e.g., an energy dashboard talking to a booking system).

- Data mapping and normalisation processes, to ensure consistency across naming conventions, units of measurement, and time intervals.

Crucially, interoperability allows smart buildings to avoid vendor lock-in and future-proof their infrastructure. It means that whether you upgrade your room booking system or deploy a new sensor network, the core data infrastructure remains intact and functional.

Real-world interoperability use cases include:

- Automatically adjusting HVAC settings based on room bookings pulled from Microsoft Exchange or Google Calendar

- Triggering security alerts if presence is detected in an un-booked and unauthorised space

- Integrating user feedback surveys with IoT sensor data to monitor satisfaction trends

Together, data tagging, semantic layering, and interoperability form a cohesive framework that turns disconnected, technical infrastructure into a living, breathing, responsive environment — one that aligns with the strategic goals of the business and adapts to human needs in real-time.

Using historical and real-time data to drive decisions

In a smart workplace, data isn't just a by-product of operations — it becomes a strategic asset. When leveraged correctly, both historical and real-time data offer powerful insights that inform decisions at every level: operational, experiential, financial, and strategic. The fusion of these two data streams — what has happened and what is happening now — empowers organisations to act proactively, not reactively, and to build environments that continually optimise themselves.

Historical Data – Learning from the Past

Historical data provides the foundation for understanding long-term patterns and trends. By analysing months or years of operational data — from space utilisation to energy consumption — organisations can uncover deep insights that would otherwise go unnoticed. These include:

- Space utilisation trends – identifying consistently underused areas that could be repurposed or downsized

- Energy efficiency benchmarking – understanding which zones or buildings are consistently over-consuming energy
- Equipment reliability analysis – using maintenance logs and sensor data to predict failure rates and plan preventive maintenance

- Occupant behaviour patterns – observing recurring patterns in room booking behaviour, desk occupancy, or collaboration zones usage

- Seasonal variations – correlating HVAC or lighting loads with external weather data to inform energy strategy

Historical data allows facilities and workplace leaders to make informed, evidence-based decisions — such as resizing real estate portfolios, investing in retrofit programmes, or redesigning user journeys. When combined with cost data, it becomes a cornerstone of return-on-investment (ROI) and total cost of ownership (TCO) models, which are crucial for justifying smart building investments.

Real-Time Data – Acting in the Now

Real-time data, on the other hand, provides immediate visibility into the dynamic state of the workplace. It enables systems to respond contextually to the needs of users and environmental conditions — often autonomously. This includes:

- Live occupancy detection – redirecting users to available desks, meeting rooms, or quiet zones

- Air quality and thermal comfort monitoring – adjusting HVAC based on thresholds for CO_2, humidity, or temperature

- Reactive maintenance – triggering alerts when a system fails or performs outside of normal bounds

- User behaviour tracking – such as dwell time near digital signage, or movement patterns within spaces, to optimise layout and flow

- Energy load balancing – adjusting lighting and HVAC in real time to meet demand without overconsumption

This real-time responsiveness creates a fluid, seamless user experience and improves operational agility. In environments where employee wellbeing, sustainability targets, or uptime are paramount, real-time insights are not just useful — they're essential.

Blending Both: The Power of Predictive and Prescriptive Intelligence

The true power of data is unlocked when historical and real-time sources are combined and analysed together using AI, machine learning, or rules-based engines. This paves the way for predictive and prescriptive intelligence:

- **Predictive analytics** – anticipating when a space is likely to be full, when systems may fail, or when energy demand will peak

- **Prescriptive analytics** – suggesting or triggering specific actions to prevent disruption, improve experience, or save cost

For example, if data shows that a certain meeting room is consistently too hot and underutilised, the system might suggest adjusting HVAC parameters, notifying FM teams, or reconfiguring that space entirely. Or, if real-time desk booking shows a surge in on-site attendance on certain days, predictive insights could prompt the business to open overflow zones automatically.

Decision-Making at Multiple Levels

Well-structured data supports decisions across three horizons:

- Operational – adjusting cleaning rotas based on usage, triggering HVAC shutoff in vacant areas, optimising space usage dynamically

- Tactical – planning for new workspace configurations, responding to employee feedback, trialling new layouts

- Strategic – shaping long-term real estate strategy, sustainability goals, or technology investment roadmaps

The 360 Smarter Stack methodology champions this layered view, ensuring that data flows not only across systems but also up through decision-making layers, giving C-suite leaders, FM teams, and IT stakeholders the confidence and clarity to act decisively.

Ultimately, combining historical and real-time data transforms the smart office from a static environment into a responsive, learning ecosystem. It's no longer just a place people work — it's a platform that understands, adapts, and evolves.

Chapter 11: From Data to Insight

As smart workplaces generate ever-growing volumes of data — from sensors, user devices, space booking apps, environmental systems, and beyond — the challenge shifts from collection to interpretation. Raw data, on its own, holds limited value. The real power lies in transforming that data into actionable insight that supports decisions, streamlines operations, and enhances the user experience.

To achieve this, organisations must implement robust analytics frameworks, leverage intuitive dashboards, and move toward predictive and even prescriptive capabilities. But this must be done responsibly — balancing innovation with clear governance over data privacy, ownership, and ethical use. This chapter explores how smart workplaces can extract meaningful insights from data while maintaining trust, transparency, and long-term value.

Analytics, Dashboards, and Predictive Insights

In the smart workplace ecosystem, the sheer volume of data generated by sensors, systems, and user interactions can be overwhelming without a structured mechanism for interpretation. This is where analytics plays a transformative role. By turning raw data into narratives — stories that explain what is happening, why it's happening, and what should happen next — organisations can make confident, evidence-based decisions that drive performance, sustainability, and user satisfaction.

From Data to Decision-Making

At its core, analytics is about clarity and confidence. It moves decision-making from instinct or tradition to a position grounded in facts, trends, and real-time visibility. The smart office, built on layers of interconnected technologies from the 360 Smarter Stack, is uniquely positioned to harness analytics at multiple levels:

- Operational optimisation – adjusting HVAC in real time to match room occupancy.

- Strategic space planning – right-sizing office space based on long-term usage trends.

- Employee experience – tailoring services or amenities based on usage and feedback data.

But for these insights to be meaningful, they must be presented in context — and that's where dashboards, visualisations, and predictive models come into play.

Dashboards: A Single Pane of Glass

Dashboards serve as the visual front-end of a smart workplace's intelligence system. They synthesise data from a wide array of sources — including booking systems, BMS, IoT sensors, energy meters, and access control — into clean, intuitive displays tailored to specific roles. For example:

- A facilities manager may view space utilisation, temperature control, and maintenance issues at a glance.

- An HR leader may track occupancy against employee wellbeing metrics.

- A sustainability officer may monitor live energy performance or CO_2 emissions.

These dashboards are increasingly interactive, allowing users to filter data by location, time period, or usage type. And critically, modern systems are role-based — meaning users only see what's relevant to their responsibilities, reducing noise and enhancing clarity.

Beyond the Dashboard: Predictive and Prescriptive Analytics

While dashboards excel at showing what is happening now or has happened in the past, predictive analytics looks ahead. By applying machine learning and statistical modelling, smart office platforms can identify likely future scenarios:

- Forecasting desk demand for hybrid teams over the coming months

- Predicting HVAC maintenance requirements based on usage patterns and fault history

- Anticipating cleaning needs based on scheduled events, occupancy spikes, or sensor triggers

Even more advanced is prescriptive analytics, where the system doesn't just forecast — it recommends action. For example:

- "Reduce lighting on Floor 3 by 15% during afternoons to save energy."

- "Reschedule maintenance to avoid disrupting high-traffic areas."

- "Open additional collaboration areas on Wednesdays based on expected team attendance."

This shift from descriptive to prescriptive enables organisations to move from simply being informed to becoming genuinely intelligent and adaptive.

Real-Time Insights and Operational Responsiveness

Real-time analytics support dynamic workplace responsiveness. Consider these scenarios:

- A meeting room is booked — the analytics engine triggers the lighting and HVAC systems in advance.

- An occupancy spike is detected — the ventilation rate is adjusted automatically to maintain comfort.

- A user reports poor air quality through a mobile app — the system cross-references air quality sensors, identifies the issue, and notifies the relevant teams.

This kind of data orchestration across systems is only possible when analytics are tightly integrated with the broader workplace infrastructure, as mapped in the 360 Smarter Stack.

Benchmarking and Continual Improvement

Analytics also support longitudinal analysis and benchmarking:

- Comparing building performance across regions

- Measuring the effectiveness of workplace changes (e.g., post-refurbishment)

- Tracking progress against ESG goals such as energy consumption or diversity of space usage

Organisations that treat data as a strategic asset — not just an operational by-product — are better positioned to continuously evolve their workplaces, meet compliance needs, and optimise investment.

Data Ownership, Privacy, and Lifecycle Considerations

In the digital workplace, data is both an opportunity and a responsibility. As organisations strive to harness data to enhance operations, boost employee experience, and deliver on ESG targets, they must also navigate complex questions about who owns the data, how it is protected, and how long it should be retained. These aren't just technical decisions — they are ethical, legal, and reputational imperatives.

Data Ownership: Who Holds the Keys?

Data generated within a smart workplace can come from multiple sources: employee interactions, sensor systems, third-party platforms, building management systems, and more. The first consideration is ownership — who has the right to access, control, or monetise that data?

- For employee data (e.g. occupancy behaviour, system usage, comfort feedback), organisations must balance operational insight with personal autonomy and trust.

- For vendor-generated data (from booking apps or integrated solutions), contractual clarity is essential. Does the vendor retain rights to use anonymised data across clients? Are there limits on where or how the data is stored?

- Facilities, IT, HR, and Security teams may each claim ownership of overlapping datasets — requiring clear data governance models to avoid conflict and duplication.

As smart office initiatives grow in complexity, it is increasingly valuable to establish a data stewardship framework — assigning responsibility for specific datasets and ensuring alignment across systems and stakeholders.

Privacy: Creating Trustworthy Workplaces

Privacy is a cornerstone of modern digital environments, particularly in regions governed by regulations like GDPR in the UK and EU. While employees may accept some level of monitoring for operational purposes, they must be assured that their personal data is handled transparently, fairly, and securely.

Key privacy considerations include:

- Anonymisation and aggregation – ensuring individual users cannot be identified in analytics outputs unless expressly required and consented to.

- Clear user consent mechanisms – particularly for mobile apps or opt-in services such as environmental feedback or personalisation features.

- Privacy-by-design – embedding safeguards into the architecture of smart systems, rather than retrofitting them as an afterthought.

Organisations should also invest in communications and awareness: explaining what data is being collected, why, and how it will (and won't) be used. This transparency builds trust and encourages adoption of workplace technologies.

Lifecycle Management: Data From Cradle to Grave

Data does not live forever. Managing the full lifecycle of workplace data — from creation to eventual deletion — is essential for security, compliance, and storage efficiency.

- Retention policies should align with the purpose of the data. For example, daily room booking logs might be stored for a few weeks, while energy usage trends may be retained for years to support ESG reporting.

- Archiving strategies can allow older, less frequently accessed data to be stored at lower cost while remaining accessible for audit or trend analysis.

- Deletion protocols must be robust, particularly for personally identifiable information (PII) or sensitive operational data.

Integrating lifecycle rules into data models and governance tools ensures that data remains both usable and compliant.

Security and Resilience

Closely linked to lifecycle and ownership is the need for robust cybersecurity. Smart offices are inherently interconnected, with multiple access points, cloud integrations, and mobile interfaces. This increases the attack surface — making security-by-design non-negotiable.

Best practices include:

- Role-based access controls to limit data visibility to authorised personnel

- Encryption at rest and in transit to protect sensitive data

- Regular audits and vulnerability assessments across platforms and integrations

Crucially, smart office platforms should be designed to fail gracefully — maintaining basic operations even if data feeds or connections are disrupted.

Embedding Governance into the 360 Smarter Stack

Data ownership, privacy, and lifecycle considerations are not standalone issues. They must be embedded at every layer of the 360 Smarter Stack:

- In system design and platform selection

- In integration architecture and API management

- In user interfaces and reporting tools

- In change management and governance processes

By treating data not just as a tool, but as a shared organisational asset, smart workplaces can create environments that are not only efficient and intelligent — but ethical, secure, and human-centred.

Layer 6: Integration Layer

Chapter 12: Getting Systems to Talk

As the smart office ecosystem expands, it brings together a growing array of technologies — HVAC systems, lighting controls, occupancy sensors, booking platforms, digital signage, energy dashboards, and more. Yet these systems are often developed by different vendors, built on different protocols, and managed by different departments. Without seamless integration, the promise of a smart workplace quickly unravels into a tangled web of siloed applications.

The ability to orchestrate interactions between systems — to enable them to "talk" to one another in real time — is what transforms a collection of tools into a unified smart environment. This chapter explores the technical foundations that make this possible: middleware, APIs, integration platforms, and open standards. It also reflects on practical patterns and hard-won lessons from real-world integration efforts.

Role of Middleware, APIs, and Integration Platforms

Creating a truly connected smart office requires far more than simply installing a series of smart devices or software tools. The real transformation occurs when these disparate components — often developed by different vendors, speaking different digital languages — can exchange data, respond to real-time events, and coordinate actions intelligently. This is made possible through an architectural backbone built on middleware, APIs, and integration platforms.

Middleware: The Translator of the Smart Office

Middleware serves as the invisible layer that enables communication between systems which would otherwise be isolated. Think of it as the translator at a multilingual conference — converting data formats, managing communication flows, and ensuring that systems which were never designed to work together can still collaborate.

In smart buildings, middleware often provides:

- Protocol translation, such as converting building-level BACnet or Modbus signals into IP-based formats readable by cloud applications

- Data harmonisation, ensuring consistency in naming conventions, units, and time stamps

- Event orchestration, for example, enabling an occupancy sensor to trigger lighting, HVAC, and room availability updates simultaneously

Middleware is especially critical when integrating legacy systems with modern cloud-based applications. Instead of replacing entire infrastructures, middleware allows organisations to layer on new functionality incrementally and cost-effectively.

Examples include:

- Spica Workplace Hub, which acts as a data aggregation and control layer

- Microsoft Dataverse, part of the Power Platform, enabling data flow across apps and services

- BuildingOS by Acuity Brands, which consolidates building performance data for actionable insights

APIs: The Rules of Engagement

Application Programming Interfaces (APIs) define how software components should interact. In smart office environments, APIs are essential to exposing data, enabling control, and establishing two-way communication between systems.

For instance:

- A space booking app might call an API to check desk availability, book a seat, and notify cleaning teams

- A digital signage solution may use an API to pull meeting data from a calendar system

- An environmental monitoring app could ingest CO_2 sensor data via API to trigger alerts or ventilation adjustments

Well-designed APIs follow open industry standards (e.g., RESTful APIs, GraphQL) and provide secure, scalable access to both data and services. They are critical for future-proofing — ensuring that systems can evolve, integrate with new tools, and support mobile or remote interfaces.

Increasingly, vendors differentiate themselves on API richness — with some offering developer portals, SDKs, and sandbox environments to support deeper customisation and innovation.

Integration Platforms: The Glue at Scale

As the number of smart office systems grows, manually connecting each one becomes impractical. That's where integration platforms — sometimes known as iPaaS (Integration Platform as a Service) — come in.

These platforms provide a centralised environment to:

- Build, monitor, and manage data flows between applications

- Trigger automations based on business logic

- Implement workflows without writing code (using drag-and-drop interfaces)

In practical terms, this means facilities, IT, or workplace teams can:

- Automatically start heating in a room when it's booked and occupied

- Push fault alerts from sensors to a helpdesk ticketing system

- Feed real-time occupancy data into Power BI dashboards

Some of the most widely used platforms in the enterprise space include:

- Microsoft Power Platform (Power Automate, Power Apps, Power BI, Dataverse)

- Cisco Spaces for spatial intelligence and IoT device orchestration

- MuleSoft for more complex, enterprise-grade integrations

- Zapier or Make.com, offering lightweight automation for SMEs

These platforms are enablers of speed and agility — allowing organisations to prototype quickly, test new use cases, and scale successful ones without heavy IT overhead.

Importance of Open Standards (MQTT, BACnet, KNX, etc.)

As smart offices grow in complexity, the ability for systems to interoperate seamlessly becomes mission critical. But integration isn't just about APIs and middleware — it's also about the language that devices and systems speak. This is where open standards come into play. These standardised protocols and frameworks ensure that different technologies — often from different manufacturers and eras — can understand each other, communicate reliably, and evolve together.

Why Open Standards Matter

In any modern smart office, there may be dozens — even hundreds — of systems involved: HVAC controls, occupancy sensors, lighting arrays, audio-visual systems, access control, elevators, and room booking apps, to name a few. Without open standards, each of these systems could end up operating in its own silo, with limited or no ability to share data or orchestrate services.

Open standards address this by:

- Reducing vendor lock-in, allowing organisations to select best-of-breed technologies without worrying about incompatibility

- Enabling plug-and-play integration, speeding up deployment and minimising custom development

- Supporting long-term scalability, so buildings can evolve without complete rewiring or replacement

- Improving resilience, because open protocols are supported by multiple vendors and communities

By using devices and platforms built on common standards, smart offices gain agility —
both in implementation and in future adaptation.

Key Open Standards in the Smart Building Ecosystem

Here are some of the most widely used open standards in smart office environments:

☐ MQTT (Message Queuing Telemetry Transport)

- A lightweight, publish-subscribe messaging protocol designed for constrained devices and low-bandwidth networks — making it ideal for IoT.

- Commonly used for connecting environmental sensors, occupancy detectors, and other field devices to cloud platforms.

- Supports real-time streaming of data and event-driven architectures.

Use case: An MQTT broker collects real-time data from temperature sensors across multiple office floors, pushing updates to dashboards and automation workflows.

☐ BACnet (Building Automation and Control Network)

- A communications protocol specifically developed for building automation systems.

- Widely used for HVAC, lighting, fire detection, and access control.

- Promotes interoperability between systems from different vendors.

Use case: BACnet enables a smart HVAC system to receive occupancy data from the building management system and adjust air handling accordingly.

☐ KNX

- A standard for home and building automation that covers everything from lighting and blinds to heating, security, and energy management.

- Often used in European smart buildings due to its robustness and flexibility.

- KNX devices can be connected via twisted pair, IP, RF, or powerline.

Use case: A KNX lighting control system adjusts ambient lighting throughout the day, based on programmed scenes and daylight availability.

☐ Modbus

- A long-standing protocol for connecting industrial electronic devices, still widely used in building management systems (BMS).

- Often employed for communication between sensors, meters, and controllers at the field level.

Use case: A Modbus energy meter reports live consumption data to a central energy dashboard, helping facilities monitor efficiency targets.

☐ OPC UA (Open Platform Communications Unified Architecture)

- A machine-to-machine communication protocol for industrial automation.

- Supports secure and reliable exchange of data in the context of industrial IoT (IIoT) and building systems.

Use case: OPC UA connects different industrial controllers and subsystems in a mixed-use building, allowing supervisory control across HVAC and production areas.

Balancing Standards with Innovation

While open standards are essential, it's also important to strike the right balance. Newer and more innovative systems may initially rely on proprietary protocols or APIs before broader standardisation catches up. The key is to build a flexible architecture that supports both standardised and custom integrations — ideally through an abstraction layer that can normalise data formats and flows.

Standards as Strategic Enablers

Adopting open standards is not just a technical decision — it's a strategic move. It enables greater choice, increases resilience, and future-proofs investments. By embedding open standards into procurement criteria and architectural blueprints, organisations ensure that their smart office initiatives remain adaptable, extensible, and aligned with the evolving digital landscape.

Real-World Integration Patterns and Lessons Learned

As organisations embark on smart office transformations, integrating multiple systems—from legacy building management systems (BMS) to modern cloud platforms—often becomes one of the most complex and critical challenges. The real world rarely mirrors the neat diagrams seen in vendor presentations. Instead, messy APIs, inconsistent data formats, siloed systems, and security concerns emerge.

In this expanded section, we explore real-world integration patterns, design approaches, and lessons learned from active deployments, highlighting how integration shapes the daily operation of a smart workplace. These patterns are aligned with the 360 Smarter Stack methodology, which promotes modularity, interoperability, and a service-oriented mindset.

1. Pattern: Event-Driven Automation via Middleware

Example: Booking a meeting room automatically activates HVAC, adjusts lighting, sends a digital concierge message to signage, and powers on AV equipment.

Expanded Insight:

- At the heart of this pattern is event-driven architecture (EDA). Events like "room booked" or "employee checked in" trigger workflows orchestrated by middleware (e.g., Node-RED, Azure Logic Apps, or n8n).

- Middleware decouples systems, meaning each service subscribes to relevant events and reacts independently, promoting resilience and scalability.

- Integrations can use message brokers (e.g., MQTT, Kafka) for real-time, low-latency updates, or webhooks for lighter interactions.

Lessons Learned:

- Context matters—a room booking event should include metadata (user type, time of day, purpose) to tailor the environment.

- Event queues must be monitored and logged, and retry mechanisms built in for resilience.

- Stakeholder alignment is essential—without consistent understanding of event outcomes, workflows may trigger unintended consequences (e.g., AV activating outside business hours).

2. Pattern: API Gateway with Unified Interface Layer

Example: A unified front-end dashboard draws data from disparate systems—air quality sensors, desk sensors, lighting controllers, visitor logs—via REST APIs.

Expanded Insight:

- An API gateway such as Microsoft Azure API Management, AWS API Gateway, or MuleSoft acts as a front door, standardising how systems interact.

- Behind the gateway, API orchestration logic aggregates, filters, and reshapes data from different sources into a consistent structure.

- Authentication and rate limiting are centrally managed, improving security and performance.

Lessons Learned:

- An early mistake is to call APIs directly from frontend apps—introduce a middleware layer to shield client interfaces from underlying complexities.

- Version control and deprecation strategies must be planned to avoid breaking dependencies when underlying systems update.

- Prioritise RESTful standards and OpenAPI specifications—consistency reduces integration friction and accelerates developer onboarding.

3. Pattern: Bridging Operational Technology (OT) and IT Systems

Example: Legacy BMS systems using BACnet or Modbus need to feed real-time energy data to a cloud-based ESG reporting tool.

Expanded Insight:

- Protocol converters (e.g., BACnet/IP to MQTT bridges) enable industrial equipment to "talk" to modern systems.

- Platforms like Niagara Framework or KMC Commander can act as middleware hubs translating and tagging data for IT systems.

- The semantic gap between raw BMS data and business logic must be bridged with metadata tagging and context-aware transformation.

Lessons Learned:

- Plan for physical access—many BMS systems are behind firewalls or air-gapped for security. VPN tunnelling or edge devices may be required.

- Align with IT cybersecurity standards when exposing OT data to the cloud—this includes encryption, role-based access, and secure transport protocols.

- Naming conventions and tagging are mission critical consistency ensures scalability and meaningful analysis.

4. Pattern: Digital Twin Integration

Example: A digital twin visualises and simulates occupancy, energy use, and indoor air quality in real time for predictive facilities planning.

Expanded Insight:

- Digital twins rely on semantic data models, such as RealEstateCore, Brick Schema, or IFC, to map building elements and assets.

- Real-time data feeds from IoT platforms (e.g., Azure Digital Twins, TwinMaker) update the twin's virtual state continuously.

- This enables "what-if" scenarios, such as predicting the effect of HVAC adjustments on comfort and energy spend.

Lessons Learned:

- Digital twins are only as good as their data fidelity—gaps or inconsistencies in IoT data can distort simulations.

- Establish an asset registry and clear hierarchy for rooms, zones, equipment, and services early in your design.

- Build incrementally—start with pilot zones or systems before extending twin models across entire estates.

5. Pattern: No-Code/Low-Code Workflow Automation

Example: Facilities managers create automated flows that alert cleaning teams when a certain threshold of hot-desk usage is reached.

Expanded Insight:

- Tools like Microsoft Power Automate, Zapier, and Tray.io empower non-developers to build automated logic across apps and data sources.

- These platforms offer pre-built connectors for Office 365, ServiceNow, IoT platforms, and even legacy email alerts.

Lessons Learned:

- Empowering business teams to automate creates agility—but without governance, it can lead to "automation sprawl."

- Introduce a review process for published flows, with documentation and monitoring.
- Consider training a team of "citizen developers" who can safely innovate within a controlled environment.

Integration Design Principles

Across all patterns, several key design principles emerge:

- Loose Coupling, Tight Context: Integrations should minimise direct dependencies but ensure data has rich context and traceability.

- Semantic Consistency: Use common naming conventions, ontologies, and metadata structures to align across platforms.

- Observability: Build in logging, error tracking, performance monitoring, and alerting to support continuous improvement.

- Security-First Mindset: Always use least privilege, encrypt in transit and at rest, and segment networks to reduce risk.

Real-world smart office integrations rarely follow a single playbook. Instead, successful organisations apply repeatable integration patterns, understand their specific context, and plan for evolution over time. Integration is not a one-off task—it's a living capability that needs care, iteration, and ownership.

In a world of rapidly shifting workplace needs, regulatory requirements, and user expectations, the ability to stitch together systems—both old and new—is what will differentiate tomorrow's truly smart workplaces.

Layer 7: Systems & Devices

Chapter 13: Deploying Smart Building Systems

Deploying smart building systems is where the strategic vision becomes tangible—where planning turns into physical installation, configuration, and operation. This chapter explores the key technologies that form the foundation of a smart office—ranging from AV and HVAC to IoT sensors and lighting—and the best practices for procuring, commissioning, and configuring these systems to ensure reliability, interoperability, and future-proofing.

The goal isn't just to install devices; it's to deploy systems in a way that integrates seamlessly with the broader workplace ecosystem and supports the contextual data model and user experience goals defined in earlier chapters. With proper orchestration, even legacy infrastructure can be uplifted into intelligent, responsive components of a dynamic workspace.

Let's begin by breaking down the core categories of systems commonly deployed in smart workplaces.

AV, HVAC, Lighting, Access Control, IoT Sensors

Modern smart offices rely on a tightly integrated web of systems that span across audiovisual communication, environmental control, workplace safety, and real-time sensing. These systems form the operational backbone of a responsive, intuitive, and data-informed workspace. Deploying them successfully requires far more than technical installation—it demands strategic planning, cross-team coordination, interoperability, and a deep understanding of both human and organisational needs.

Let's examine each of these system categories in greater detail:

Audio Visual (AV) Systems

AV is the most visible expression of a smart workspace and is often the first point of engagement for hybrid teams. Whether it's a huddle room, an executive boardroom, or a multi-purpose event space, the expectations are high—clarity, reliability, and ease of use are non-negotiable.

Key AV deployment considerations include:

- Unified Control: Touch panels, voice activation, or mobile apps that allow users to control lighting, screens, blinds, and conferencing in one interface.

- Wireless Collaboration: Native support for screen sharing from any device (BYOD), with zero configuration friction.

- Hybrid Meeting Enablement: AI-powered auto-framing cameras, active speaker tracking, spatial audio, and real-time transcription elevate the virtual meeting experience.

- Device Management Platforms: Tools like Crestron XiO Cloud or Logitech Sync provide remote diagnostics, firmware management, usage analytics, and proactive support alerts.

AV deployments should be standardised across locations where possible, with emphasis on user onboarding, room layout optimisation, network segmentation, and AV over IP readiness to future-proof the investment.

HVAC (Heating, Ventilation, and Air Conditioning)

Traditionally invisible to the user, HVAC now plays a starring role in the smart office due to its close ties to wellbeing, energy consumption, and ESG reporting.

Modern smart HVAC implementations include:

- Occupancy-Driven Airflow: Automatically adjusts temperature and ventilation levels based on room usage in real time, enhancing both comfort and energy efficiency.

- Air Quality Monitoring: Integration with CO_2, VOC, humidity, and particulate sensors enables automatic modulation to maintain healthy indoor air conditions.

- Demand-Controlled Ventilation (DCV): Reduces fan speed or fresh air intake when areas are unoccupied—balancing cost, performance, and sustainability.

- Predictive Maintenance: Using sensors and historical data to detect anomalies in airflow, refrigerant levels, or filter status before system failure.

Integration is often achieved through BACnet/IP, Modbus, or MQTT, enabling HVAC systems to become part of the central building data platform.

Lighting Systems

Lighting shapes mood, focus, and alertness. Smart lighting does more than simply switch on and off—it responds dynamically to environment, occupancy, and even individual preference.

Features of a smart lighting system include:

- Human-Centric Lighting (HCL): Adjusts colour temperature and intensity throughout the day to mimic natural daylight rhythms and reduce fatigue.

- Zoned and Task-Based Control: Enables specific areas to respond independently based on schedules, occupancy sensors, or user activity.

- Daylight Harvesting: Uses ambient light sensors to adjust artificial lighting levels, reducing energy consumption while maintaining optimal visibility.

- DALI-2 and KNX Standards: These protocols enable modular, interoperable deployments and allow easy integration with other building systems (e.g. AV, HVAC, fire alarms).

Lighting design must account for emergency lighting compliance, user overrides, maintenance access, and power backup for critical areas.

Access Control Systems

In the smart workplace, access control serves both physical security and operational efficiency. Gone are the days of plastic cards and siloed systems—today's access platforms are intelligent, data-rich, and highly customisable.

Modern features include:

- Mobile Credentialing: Employees can unlock doors using smartphones, wearables, or biometric identifiers, often through integration with corporate identity systems.

- Zone-Based Access and Scheduling: Permissions can vary by team, time of day, or building location—enabling more granular control and better safeguarding of sensitive areas.

- Integration with Booking Systems: Ensures that only those with an active desk or meeting room reservation can gain entry to a space, improving space utilisation and security.

- Visitor Management: Seamless digital check-in, badge printing, and notification systems streamline guest access while maintaining safety protocols.

Many modern access control systems expose REST APIs, Webhooks, or support open standards like OSDP, enabling them to feed data into security dashboards, visitor analytics, or wider automation workflows.

IoT Sensors

The nervous system of any smart building, IoT sensors provide the live data streams that enable responsiveness, analytics, and automation.

Core categories include:

- Occupancy and People Counting: Infrared, ultrasonic, or camera-based sensors detect presence or count individuals in real time—feeding into HVAC, cleaning, booking, and safety systems.
- Environmental Monitoring: Measures air quality (CO_2, VOCs, PM2.5), temperature, humidity, and noise levels. These insights can be visualised in wellbeing dashboards or used to trigger HVAC or lighting changes.

- Desk and Room Sensors: Monitor usage patterns, detect abandonment, or enable touchless check-ins.

- Smart Utilities and Energy Meters: Capture real-time power, gas, or water usage at a granular level, helping identify inefficiencies or support sustainability initiatives.

- Maintenance Sensors: Vibration or fluid level sensors on critical infrastructure (e.g., pumps, generators) can flag wear-and-tear, enabling condition-based maintenance.

Deployment requires careful planning around network type (LoRaWAN, Zigbee, BLE), battery life, data retention policies, and integration with middleware.

Procurement, Commissioning, and Configuration Best Practices

The deployment of smart building technologies is a major investment—not just in capital, but in culture, process, and digital maturity. Far too often, organisations fall into the trap of treating the selection and roll-out of these systems like any other IT project or FM upgrade. In reality, smart building initiatives require a holistic, lifecycle-driven approach—beginning at procurement, reinforced through commissioning, and sustained through well-governed configuration.

Getting these stages right means not only ensuring the technology works but that it works well—for your people, your spaces, and your long-term ambitions. It's about establishing the foundations for intelligent, adaptive, and truly human-centric workplaces.

Smart Procurement: Shifting from Product to Platform Thinking

Procurement is often the first critical point of failure—or success. The complexity of smart buildings means buyers need to look far beyond isolated features or the cheapest supplier. Instead, they should be evaluating ecosystem fit, scalability, and strategic alignment.

Key principles for smart procurement:

- Define Capabilities, Not Just Specs: Move beyond checklists. Describe the outcomes you want—automated room experiences, improved energy visibility, reduced service calls—so vendors are forced to respond with relevant solutions, not generic brochures.

- Procure Platforms, Not Point Products: Wherever possible, invest in flexible, modular platforms (e.g., IWMS, middleware hubs, or experience suites) that can evolve. Point products may solve a need now but become digital dead ends later.

- Weight Interoperability Highly: Systems that can communicate—via APIs, standard protocols, or event hubs—unlock value over time. Ask vendors to demonstrate integrations already working with your systems or ecosystem (e.g., Microsoft, Cisco, Johnson Controls).

- Vendor Maturity and Roadmap: A flashy interface doesn't mean a strong partner. Investigate the vendor's financial stability, update frequency, partner network, and approach to cybersecurity. Will they still be here (and relevant) in five years?

- Proof of Concept Contracts: For emerging or strategic technologies, consider trial deployments before signing multi-year agreements. This reduces long-term risk and gives your team real-world exposure to new systems.

- Soft Factors Matter: Assess support quality, training materials, deployment experience, and cultural fit. A technically capable vendor that doesn't communicate well or respect change management processes may cause more pain than they solve.

Commissioning: Beyond Technical Sign-Off

Commissioning is often misunderstood as a box-ticking exercise—but in the smart building context, it's a vital opportunity to validate user experiences, fine-tune integrations, and establish a stable foundation for operations.

Best practices for commissioning include:

- Integrated Use Case Testing: Don't just test whether the lights turn on. Simulate real-world behaviours—like a user booking a room, arriving with a guest, using AV equipment, and leaving early. Observe how systems respond across the chain.

- Cross-Disciplinary Participation: Bring together IT, FM, workplace experience, security, and even end users. Commissioning is a chance to align understanding, raise final issues, and transfer ownership across departments.

- 'Day in the Life' Scenarios: Create scenario scripts that mimic entire workflows. Example: a visitor journey from invite to reception check-in to space access to departure. This holistic view helps catch integration gaps and user experience hiccups.

- Environmental Verification: For systems like HVAC or lighting, verify sensor placements, daylight response, comfort ranges, and noise levels—not just digital settings.

- Cybersecurity and Network Readiness: Include tests for firmware currency, encryption settings, secure onboarding (e.g., 802.1X), segmentation, and resilience. Often neglected, these factors are vital in converged systems.
- Create a Digital Commissioning Record: Document settings, firmware versions, IP schemes, API keys, test outcomes, open issues, and asset metadata. Use cloud-hosted tools for ongoing access and version tracking.

Training as Part of Commissioning: Ensure that the handover process includes technical onboarding, admin role access, support contacts, and issue escalation paths. This avoids 'go live and disappear' scenarios.

Configuration: Engineering for Scalability, Usability, and Insight

Configuration is where the potential of a smart system is either unlocked—or undermined. Far more than ticking options in a UI, configuration should be seen as an act of digital architecture: shaping how people interact, how data is generated, and how the building responds.

Detailed best practices:

- Create and Enforce Naming Conventions: Establish consistent, logical naming for devices, rooms, locations, and users (e.g., Floor10-East-WC01 vs "Room1"). This clarity flows into analytics, reporting, and helpdesk logs.

- Leverage Data Tagging and Semantic Metadata: Assign clear tags and metadata to every sensor, asset, or stream. For example: Zone: NorthWing, Type: CO_2, AssetClass: HVAC, PowerSource: Battery. This enables filtering, analytics, and cross-platform context.

- Configure With the User in Mind: Avoid technician-centric defaults. For example, set meeting room touch panels to show plain-English room names, logical calendar views, and minimal taps. Hide advanced settings from general users.

- Role-Based Access and Auditing: Use permissions to control who can see or edit what. Ensure logs are retained for changes. Especially in multi-tenant or high-security buildings, this is non-negotiable.

- Configure Alerts Intelligently: Define not only thresholds, but when and how alerts are triggered. Use escalation rules, suppression schedules (e.g., during maintenance), and contextual messaging. Example: "CO_2 threshold exceeded in Zone C – meeting rooms occupied at 95%."

- Version Control Configuration: Where possible, keep snapshots of configuration states or exportable settings files. This helps with recovery, duplication, and auditing over time.

- Integration-Ready Outputs: Ensure that any configured outputs—like presence data, energy readings, or booking states—are exposed via API, MQTT, or webhook. This allows orchestration platforms or IWMS layers to consume and act on them.

- Continuous Configuration Culture: Don't treat configuration as a one-time activity. Build in periodic reviews (e.g., every 3–6 months) to adjust settings based on feedback, space changes, or new building usage patterns.

By approaching procurement, commissioning, and configuration not as isolated project tasks, but as integrated pillars of a smart workplace strategy, organisations can avoid costly missteps and drive genuine value. These stages form the connective tissue between ambition and reality—ensuring the smart building delivers not just functionality, but resilience, insight, and delight.

Chapter 14: Reliability, Monitoring, and Maintenance

In the rush to deploy innovative smart building systems, it's easy to overlook a fundamental truth: if the technology doesn't work reliably, people stop using it. A failed booking panel, an unresponsive sensor, or a flickering AV display may seem minor—but when such issues compound, they erode trust, increase service costs, and ultimately diminish the value of your entire investment.

Reliability, therefore, must be designed into smart workplaces from the start. It's not just about system uptime, but about anticipating failures, reducing response times, and using real-time data to proactively maintain building performance. As smart offices become more dynamic and data-rich, so too must the maintenance strategies that support them.

This chapter focuses on the infrastructure, practices, and monitoring models that enable always-on, frictionless building experiences—delivered quietly in the background but essential to the success of every user interaction.

Building System Resilience and Uptime

Smart buildings—by design—are dynamic, software-enabled environments. However, this intelligence brings complexity, and complexity increases the surface area for potential failures. Building system resilience is therefore about ensuring not just uptime, but predictable performance under a range of operating conditions, from peak demand to unexpected disruptions.

1. Architectural Resilience

- ## Redundant Network Infrastructure

  Smart buildings are increasingly reliant on IP-based systems. Ensure critical services (like security and access control) have redundant network paths and failover routing between switches or VLANs. Use industrial-grade hardware in high-availability configurations where possible.

- ## System Isolation for Critical Services

  Segment building systems logically and physically. For example, access control and fire safety should operate independently of user-facing digital signage or Wi-Fi. This protects safety-critical systems from performance degradation due to non-essential services.

- ## Local Autonomy via Edge Devices

  Empower building-edge controllers to maintain basic functionality independently. HVAC control panels, occupancy sensors, and lighting controllers should have the ability to run on predefined rulesets even if central command platforms or cloud systems are temporarily unavailable.

- ## Environmental Protections

  Install core controllers and networking gear in well-ventilated, temperature-controlled, and secure enclosures. Use UPS systems with intelligent monitoring to ensure graceful shutdowns during prolonged power failures and automatic restart post-recovery.

2. Software and Cloud Resilience

- ### High Availability Cloud Services

  Choose vendors with proven resilience strategies: geographically distributed servers, automatic scaling, and failover policies. Request documented uptime metrics (e.g., 99.9% SLAs) and review any outage history for patterns.

- ### Automatic Recovery and Self-Healing

  Select systems that can attempt self-repair operations—such as reconnecting to cloud services, rebooting local devices, or resyncing data—without human intervention.

- ### Scheduled Restart Windows

  For platforms known to develop memory leaks or performance drift over time (e.g., older AV panels or third-party software bridges), implement controlled restart windows during off-hours to preserve operational reliability.

3. Organisational Resilience

- ### Clear Support Ownership

  Define responsibility boundaries between IT, Facilities, AV, and vendors. This avoids finger-pointing during incidents and streamlines recovery efforts.

- ### Standard Operating Procedures (SOPs)

  Maintain SOPs for diagnosing key systems, contacting vendors, and performing basic on-site resets. Include schematics, IP addressing, and login credentials for authorised staff in secure documentation portals.

- ### Resilience Drills

Just as fire drills test evacuation readiness, run occasional "blackout" or partial failure simulations. This uncovers gaps in staff response, communications flow, and fallback processes before a real incident occurs.

Health Monitoring and Remote Diagnostics

Ensuring that a smart building remains truly smart over time means going beyond uptime—it requires visibility, insight, and proactive action. Health monitoring turns the built environment into a self-reporting ecosystem, while remote diagnostics reduces friction between detection and resolution.

1. Monitoring Building Health

- **Device Telemetry and Status Dashboards**

  Collect and display key operational data from all connected assets. This includes online/offline status, signal strength, last sync time, firmware version, sensor readings (e.g., temperature, CO_2), and usage logs.

- **Cross-System Health Aggregation**

  Use middleware or integration platforms (e.g., Cisco Spaces, Spica, or Niagara Framework) to consolidate telemetry across different vendor ecosystems—lighting, HVAC, AV, BMS—into a single pane of glass.

- **Proactive Alerts and Fault Prediction**

  Leverage AI-driven alerts based on usage anomalies: for example, a sensor that normally triggers every hour hasn't activated in 24 hours. These early indicators can prevent a small issue from cascading into a major failure.

- **Digital Twin Alignment**

Integrate live health data into a digital twin model of the building to visually represent problem areas. For example, highlight temperature anomalies in red or show offline devices as greyed-out on a spatial floorplan.

2. Remote Diagnostics and Intervention

- **Secure Remote Access Protocols**

  Use tools like VPN tunnels, secure APIs, or edge management platforms to enable certified engineers to access and troubleshoot systems off-site. Ensure strict identity controls and audit trails.

- **System Logs and Historical Playback**

  Preserve logs of interactions, commands, and system behaviour (e.g., when a door failed to unlock or why a light failed to dim). Historical playback is essential for root cause analysis, especially in complex or multi-layered issues.

- **Automated Remediation Scripts**

  Some platforms support no-code remediation scripts. For example, "If this display is offline for more than 10 minutes, reboot it once. If still offline, raise a Level 2 incident."

- **Integration with CAFM/IWMS Systems**

  Push alerts and diagnostics directly into computer-aided facilities management (CAFM) tools or IWMS platforms. This allows automated job creation, SLA tracking, and assignment to the correct engineer or support partner.

3. Performance-Based Maintenance

- **Usage-Driven Servicing**

Rather than scheduling maintenance every 3 or 6 months, tie servicing intervals to actual system use. A meeting room with 80% utilisation may need more frequent projector lens cleaning than one used twice a week.

- **Degradation Detection**

  Identify trends such as gradually dimming light outputs, reduced signal response times, or slowly rising error counts. These subtle trends often precede full failure but go unnoticed without analytical overlays.

- **Sustainability Integration**

  Link equipment health data with environmental metrics. For example, a faulty HVAC unit consuming more energy than usual may be not just a service issue but a sustainability risk.

Layer 8: Physical Environment

Chapter 15: Infrastructure Foundations

Before a building can be considered "smart," it must be **ready to connect, power, and support** the systems that underpin intelligence—physically, digitally, and operationally. The foundation of every smart office is its infrastructure: the structured cabling, power distribution, network racks, patching strategy, and converged services that carry data and deliver energy.

Often overlooked or treated as a "commodity," the physical environment can make or break the long-term scalability, maintainability, and performance of your smart infrastructure. Whether deploying IoT sensors, advanced AV, or energy management systems, everything relies on a rock-solid base.

Let's explore Layer 8, the physical environment and determine what must be in place to future-proof your space.

Cabling, Power, Racks, Patching, PoE

Building the Nervous System of the Smart Office

In any smart building, the physical infrastructure forms the nervous system through which data, power, and control signals travel. Every sensor, display, access control point, or AV system is only as reliable as the connectivity that supports it. Neglecting physical layer design introduces risk, bottlenecks, and rework — all of which impact performance and uptime.

Let's unpack each of these foundational components to understand how they support a scalable, integrated smart workplace environment.

1. Structured Cabling: The Backbone of Connectivity

A smart office requires far more than basic network access. Systems must accommodate hundreds — or even thousands — of connected devices across multiple floors and locations. These include:

- IP security cameras

- AV-over-IP solutions

- Digital signage

- Wireless access points

- IoT sensors (temperature, motion, CO_2, humidity)

- Access control readers

- Smart lighting and PoE switches

To support this scale and diversity, structured cabling must be:

- Category 6A or better: To support 10Gbps speeds and PoE++ across distances up to 100 metres.

- Shielded (F/UTP or S/FTP): In electrically noisy environments, especially in ceiling voids or risers.

- Planned with growth in mind: Allow 20–30% overprovisioning in cable trays and IDF rooms.

Smart design principles also include:

- Zoned cabling topologies: To segment cabling by function (AV, lighting, security, Wi-Fi).

- Cable trays and containment: Segregate power and data cabling, and allow easy access for future additions.

- Documented routes and endpoints: Use floor plans and digital asset management tools to track every drop.

A well-executed cabling strategy supports faster deployment, easier maintenance, and long-term adaptability.

2. Power Provisioning: Standard and Smart Power Delivery

Every connected device needs power, but in a smart environment, it must be delivered intelligently and efficiently.

Standard Power (AC):

- Ensure diverse ring mains and clean power to AV-heavy or digital-intensive zones.

- Use RCBO protection for sensitive equipment like AV racks and PoE switches.

- In larger systems, consider dedicated circuits for security, AV, and IT, each with UPS backup.

Power over Ethernet (PoE):

PoE has become the de facto standard for powering smart devices. The latest standard — PoE++ (IEEE 802.3bt) — can deliver up to 90W per port, enabling even more devices to be deployed without separate AC power.

Use cases for PoE include:

- Smart lighting (tuneable LED panels powered and controlled via Ethernet)

- Meeting room scheduling panels

- AV equipment like cameras and audio processors

- Occupancy and environmental sensors

- Digital signage displays

- Door controllers and badge readers

Design Considerations:

- Choose enterprise-grade switches with PoE budgeting per port and total power pool.

- Segment PoE loads across multiple switches to reduce risk of full-failure events.

- Integrate with intelligent power monitoring systems to manage loads dynamically and detect anomalies.

3. Equipment Racks and Room Layouts: Optimising the Physical Core

Racks serve as the central nodes of connectivity — yet many comms rooms remain an afterthought.

A high-performing smart workplace requires:

- Standard 42U racks with airflow control (blanking plates, vented doors)

- Proper vertical and horizontal cable management to prevent tangling and strain

- Secure PDUs (preferably with IP monitoring and per-outlet control)

- Environmental sensors (temperature, humidity, door open alerts)

In larger installations, it's also common to deploy:

- Top-of-rack patch panels for floor-by-floor consistency

- Rack-mounted PoE switches with dual power feeds

- Rack-specific asset labels and digital documentation (QR-coded equipment ID)

Cooling and Environment:

- Maintain room temperatures between 18–22°C, with clear airflow paths and active monitoring.

- Isolate heat-producing equipment and avoid stacking non-passive devices in unventilated spaces.

4. Patching and Cable Management: Cleanliness Is Control

A well-patched rack saves hours of troubleshooting — and contributes directly to system uptime and ease of MACs (Moves, Adds, and Changes).

Best Practices Include:

- Colour-coded patch cables for different services (e.g., red = access control, blue = AV, green = BMS)

- Short, exact-length patch leads to prevent looping and clutter

- Patch documentation software to log connections, endpoints, and device MAC addresses

- Physical separation of PoE cabling from high-draw power or analogue signal wires

In larger estates, patching should follow a documented convention, such as:

Mathematica

- Copy
- Edit
- Rack # - Panel Port - Service - Floor
- E.g., Rack02-P14-Security-Floor5

This allows engineers to remotely diagnose issues, reduces mispatching risk, and supports training and scaling.

5. Power Over Ethernet (PoE): Smart Power at Scale

PoE has become a game-changer for smart offices. Rather than requiring localised mains power, PoE allows both power and data to travel over a single Ethernet cable — simplifying deployments and reducing cost.

Types of PoE:

PoE Type	Standard	Max Power to Device
PoE (Type 1)	IEEE 802.3af	15.4W
PoE+ (Type 2)	IEEE 802.3at	25.5W
PoE++ (Type 3 & 4)	IEEE 802.3bt	60W / 90W

Design Guidance:

- Plan PoE switch budgets per floor or per comms room based on connected load.

- Use structured power planning: no more than 60–70% average load to allow burst capacity.

- Track device wattage — a single pan-tilt-zoom camera may consume 30–50W.

Emerging Use Cases for PoE:

- Smart blinds

- PoE-powered HVAC control panels

- Meeting room booking displays with occupancy indicators

- USB-C charging hubs for desks

- AI edge devices (e.g., machine vision systems)

Cabling, power, and rack strategies may not be glamorous, but they are mission-critical to smart building success. A minor oversight — like overloaded PoE switches or poor rack airflow — can lead to failures, outages, or costly retrofits.

By investing early in robust infrastructure and following structured design and documentation practices, organisations create a smart foundation that can adapt, scale, and evolve in tandem with their workplace technology ambitions.

INFRASTRUCTURE FOUNDATIONS

CABLING	POWER	RACKS	PATCHING & PoE

- Category 6A or Better
- Shielded Cabling
- Zoned Cabling Topologies
- Documented Routes & Endpoints

- Diverse Ring Mains
- UPS Backup
- Dedicated Circuits
- PoE++ Deployment

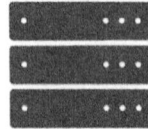

- 42U Racks Standard
- PDU & Cooling Control
- Cable Management

- Colour-Coded Cables
- Patch Documentation
- PoE Power Budgets
- Exact-Length Leads

Diagram: Infrastructure Foundations

Preparing M&E and IT infrastructure for smart capabilities

Delivering a truly smart office is not just about selecting the right digital tools and user interfaces—it's about building on a solid, integrated foundation of mechanical, electrical (M&E), and IT infrastructure. These foundational systems are the nervous system of a smart building, enabling devices, systems, and platforms to interoperate, automate, and intelligently respond to user and environmental inputs.

M&E Readiness: A Foundation for Integration

Mechanical and electrical systems such as HVAC, lighting, fire detection, lifts, water management, and backup power need to be designed with smart capability in mind from the outset. Legacy systems often operate in silos, using proprietary protocols and lacking external data exchange. Preparing M&E systems for smart enablement means:

- Open Protocol Support: Specifying building management systems (BMS) and devices that support open standards like BACnet, Modbus, or KNX enables broader integration.

- Sensor Readiness: Ensuring HVAC and lighting systems can connect with occupancy sensors, CO_2 monitors, and daylight sensors allows for automated, demand-driven control strategies.

- Addressable Systems: Systems should be deployed in an addressable manner—whether lighting circuits, power distribution units (PDUs), or fan coils—so they can be monitored and controlled individually or in zones.

IT Infrastructure: The Digital Backbone

A robust IT infrastructure is the bridge between building systems and smart functionality. To prepare for smart capabilities:

- Edge Network Architecture: Smart buildings often rely on edge computing to manage latency-sensitive or mission-critical applications. Switches and routers should be selected with this in mind, providing both data processing and analytics close to the source.

- Bandwidth Planning: With video analytics, IoT telemetry, and cloud connectivity, the network must be designed to handle high data volumes with minimal congestion, particularly in areas like meeting spaces or equipment rooms.

- Resilient Topology: Redundant connections, failover switches, and high-availability clusters ensure building operations are maintained in the event of outages or hardware failures.

- Cybersecurity Controls: As more operational technology (OT) systems are connected to IT networks, security protocols like segmentation, role-based access, and encrypted communications are essential to protect against breaches.

Coordinated Delivery

One of the most common pitfalls in delivering smart office capabilities is the misalignment between construction, M&E contractors, and IT teams. To mitigate this, projects must:

- Adopt a Common Integration Plan: A unified design document that outlines how M&E and IT systems will interface—down to protocol-level detail—ensures all disciplines work toward the same outcome.

- Introduce Digital Commissioning: Alongside traditional commissioning, teams should verify data availability, sensor calibration, control logic, and interoperability with middleware platforms.

- Sequence Smart Systems in the Build Programme: Often, the commissioning of smart systems is left too late in the programme. Smart readiness should be phased into construction milestones and not treated as a last-minute bolt-on.

Powering the Ecosystem

Finally, energy infrastructure must not be overlooked. Many smart devices—displays, IoT sensors, and even some AV equipment—can be powered over Ethernet (PoE), reducing the need for separate power cabling. This requires:

- Sufficient PoE Budgeting: Ensure switches are sized to handle total power loads across all connected devices.

- Smart Power Management: Using intelligent PDUs and UPS units allows for device-level control, load shedding, and real-time power monitoring, all feeding into the central building platform.

In summary, preparing M&E and IT infrastructure is a critical enabler for smart office ambitions. It requires a future-ready design, cross-discipline coordination, and an openness to digital convergence from day one. When done correctly, it lays the groundwork for agility, integration, and scalable transformation.

Chapter 16: Retrofitting Legacy Offices

While new builds offer the luxury of designing smart infrastructure from the ground up, the vast majority of today's workplaces are existing buildings—many with legacy systems, physical limitations, and deeply embedded operational routines. Retrofitting these environments to support modern smart capabilities requires a tailored approach: one that respects the constraints of the past while introducing scalable, future-facing technologies.

Retrofitting is not about forcing a building to become "smart" overnight, but about carefully layering in capability, improving data access, and reimagining workflows to align with what's already there. It's also about delivering incremental wins—enhancing comfort, sustainability, and operational visibility—without disrupting the building's core function or introducing excessive cost.

Constraints and Creative Solutions for Existing Buildings

Retrofitting legacy offices is a nuanced process, shaped by a mixture of physical, technical, financial, and cultural constraints. These constraints are not insurmountable—they present opportunities to innovate, adapt, and apply modern solutions in imaginative ways. Understanding these challenges and leveraging creative design and technology strategies allows older buildings to compete with their newer counterparts in functionality, user experience, and sustainability.

1. Physical Constraints: Navigating the Building Fabric

Older buildings were not designed with smart technologies in mind. Architects and engineers of the past didn't anticipate today's need for sensors, PoE networks, cloud connectivity, or occupant experience tools. As a result, we're often working within the constraints of low ceiling voids, solid walls, and architectural heritage requirements.

Challenges:

- Limited space for new wiring or ducting

- Lack of raised floors or suspended ceilings

- Preservation rules that restrict alterations

- Complex or undocumented layouts

Creative Solutions:

- Wireless IoT Solutions: Use battery-operated sensors (for occupancy, temperature, air quality) that communicate via Zigbee, Z-Wave, or LoRaWAN, bypassing the need for extensive cabling.

- Surface-Mount Installations: Cable trays, trunking, or adhesive-mounted conduits can preserve the building's fabric while adding network and power routes.

- Modular Smart Furniture: Embed sensors or displays within desks, lighting panels, or wall fixtures to avoid building-wide modifications.

- Discreet Aesthetic Design: Use architectural camouflage (e.g., colour-matched devices, embedded tech panels) to maintain heritage or listed building compliance.

2. Legacy Systems and Incompatible Technologies

Many older workplaces rely on outdated Building Management Systems (BMS), analogue HVAC controls, or access control systems built on closed protocols. These often lack APIs, modern drivers, or integration options with cloud platforms.

Challenges:

- Closed systems that don't talk to new platforms

- Limited vendor support or documentation

- High costs associated with full system replacement

Creative Solutions:

- Protocol Gateways: Devices that translate older protocols like BACnet, Modbus, or KNX into MQTT or RESTful APIs can bridge the old and the new.

- Non-Invasive Layering: Add standalone digital layers (like smart lighting or environmental monitoring) that coexist alongside legacy BMS without replacing it.

- Cloud Synchronisation: Use data mirroring or dashboard overlays to surface insights from older systems in a unified cloud interface.

- Vendor-Agnostic Middleware: Platforms like Niagara or FNT bridge disparate systems by standardising data and controls through a central integration layer.

3. Power and Network Infrastructure Gaps

Many buildings predate structured cabling, PoE (Power over Ethernet), or distributed networking closets. Electrical boards may be at capacity, and power distribution may not support the dense requirements of smart endpoints.

Challenges:

- Insufficient network drops or switch ports

- No centralised power management

- Overloaded circuits or outdated power supply

Creative Solutions:

- Localised Edge Networks: Deploy micro-networks using compact PoE switches, LTE gateways, or wireless backhaul to connect smart devices.

- PoE Extenders and Injectors: Extend the reach of power and data over existing Cat5e/6 cabling for devices like cameras, sensors, or access points.

- Solar or Energy-Harvesting Sensors: For windows, signage, and environmental sensing, use solar-powered or kinetic-powered devices.

- Smart Power Management: Introduce connected PDUs (Power Distribution Units) that can monitor and control device power consumption remotely.

4. Budgetary and Ownership Constraints

Retrofitting smart systems into leased or cost-sensitive buildings can be difficult, particularly when the long-term ROI may not benefit the current stakeholders or tenancy is uncertain.

Challenges:

- Limited capital expenditure budgets

- Reluctance to invest in long-term systems

- Landlord vs tenant conflict over responsibility

Creative Solutions:

- OPEX-based Solutions: Adopt subscription or "as-a-service" models—e.g., lighting-as-a-service, room booking platforms, or digital signage systems—that shift costs to operational budgets.

- Portable Tech Investments: Choose solutions (like occupancy sensors, displays, kiosks) that can be re-used or moved to other sites if needed.

- Energy-Saving ROI Cases: Use energy dashboards and analytics to prove the savings and payback period of proposed investments.

- Dual-Use Infrastructure: Combine smart capabilities with essential upgrades—e.g., upgrading to LED lighting with built-in sensors—to align smart goals with needed lifecycle renewals.

5. Organisational and Cultural Resistance

It's not just the building that's legacy—processes, policies, and user behaviour often lag behind too. People may be used to "how things have always worked," especially in environments where space booking or environmental controls have been static for decades.

Challenges:

- Staff reluctance to use new interfaces or apps

- Lack of digital skills among facilities or maintenance staff

- Mistrust of automation or AI-driven decision-making

Creative Solutions:

- 360 Smarter Stack Awareness: Introduce stakeholders to frameworks like the 360 Smarter Stack Methodology to build cross-functional understanding of how tech layers interact.

- Incremental Adoption: Start with intuitive, visible wins—e.g., mobile room booking, real-time meeting availability screens, or automated climate comfort.

- Change Champions: Empower early adopters or digital advocates within teams to help evangelise the benefits of smart capabilities.

- Shadow Mode Pilots: Allow new systems to run in parallel with existing ones (without replacing them) to build trust and familiarity.

6. Environmental and Sustainability Considerations

Many retrofit projects are driven by ESG goals or Net Zero commitments, but older buildings may be poorly insulated, over-lit, or reliant on carbon-heavy heating and cooling.

Challenges:

- Inability to meet sustainability benchmarks without deep retrofits

- Lack of data to inform energy use or occupancy

Creative Solutions:

- Deploy IoT Energy Monitors: Gain visibility of energy consumption by zone, floor, or asset to guide behaviour change and future investment.

- Install Smart Controls: Enable granular, automated control of lighting, HVAC, and blinds to avoid energy waste.

- Use Data for ESG Reporting: Even partial data from retrofitted systems can support compliance and help drive investment justification.

Legacy buildings hold character, history, and embedded value. Retrofitting them for smart capabilities is a journey of adaptation—working around and with constraints rather than bulldozing over them. Through thoughtful layering, edge integrations, and hybrid approaches, even the most analogue environment can become a responsive, efficient, and intelligent workplace, paving the way for a more sustainable and connected future.

Case Studies of Adaptive Transformation

Retrofitting legacy offices doesn't just involve overcoming constraints—it's also an opportunity to demonstrate how creative design, technology layering, and smart thinking can transform tired environments into modern, engaging workplaces. Across the UK and beyond, numerous examples show that adaptive transformation is not only feasible, but often more sustainable and cost-effective than new construction.

Below are a series of illustrative case studies showcasing how real organisations have embraced smart building principles to breathe new life into ageing offices.

Case Study 1: From Grey Box to Smart Hub — A 1980s Office Reborn

Location: Birmingham, UK

Building Age: 1985

Challenge: An underutilised five-storey office block with outdated HVAC, poor lighting, and no structured IT infrastructure.

Goals: Improve employee comfort, enable hybrid working, reduce energy consumption by 25%, and integrate modern AV.

Transformation:

- Smart Lighting and Sensors: A retrofit using PoE-based lighting with integrated occupancy and daylight sensors allowed for zone-level control and energy monitoring.

- Room Booking Integration: A mobile-friendly space booking system was implemented, linking desk and room availability to occupancy data.

- AV and Collaboration Upgrade: Teams and Zoom-compatible meeting rooms were created with smart occupancy-triggered AV setups.

Results: Energy use reduced by 30%, and space utilisation improved through hot-desking. The building became a testbed for future smart rollouts.

Key Takeaway: Retrofitting with intelligent lighting and modular AV created an adaptable, scalable model for modern workplace needs, without major structural changes.

Case Study 2: Listed but Not Left Behind — Smart Adaptation in a Heritage Site

Location: Edinburgh, UK

Building Age: 1890 (Grade B listed)

Challenge: Modernise office functionality without altering the protected interiors or exterior fabric.

Transformation:

- Wireless Sensor Network: A mesh of Zigbee sensors provided insights on air quality, occupancy, and light levels, requiring no intrusive cabling.

- Portable Control Hubs: Mobile control stations on trolleys offered AV, environmental, and lighting control via tablet interfaces.

- Discreet Digital Signage: E-ink displays and adaptive wayfinding were installed with reversible mounting, avoiding wall penetrations.

Results: The office retained its historic character while gaining digital wayfinding, booking, and environmental optimisation capabilities.

Key Takeaway: With creative mounting, wireless tech, and respect for the building's constraints, even listed spaces can become digitally capable without violating preservation laws.

Case Study 3: Smart on a Budget — Small Council Office Upgrade

Location: Barnet, London

Building Age: Early 2000s

Challenge: No centralised BMS, poor environmental control, and budgetary restrictions limited traditional upgrade options.

Transformation:

- Low-Cost Smart Devices: Off-the-shelf IoT sensors tracked CO_2, temperature, and occupancy in key areas like meeting rooms and open plan zones.
- Microsoft Power Platform Integration: Data from sensors fed into a Power BI dashboard, visualising comfort metrics and space utilisation in real-time.

- BYOD and Hot Desking Model: Staff were encouraged to use a desk-booking platform and report discomfort digitally via Power Apps.

Results: Improved comfort and data-driven maintenance decisions, with 60% of improvements driven by employee feedback loops.

Key Takeaway: By using existing tools (Microsoft stack) and consumer-grade smart tech, even low-budget retrofits can deliver meaningful, measurable improvements.

Case Study 4: Multitenant Smart Building Overlay

Location: Manchester, UK

Building Age: 1990s office block, multiple tenants

Challenge: No consistent infrastructure across floors; landlords couldn't enforce unified systems.

Transformation:

- Platform-Agnostic Middleware: Implemented a neutral data layer (Niagara Framework) that aggregated feeds from different tenants' systems.

- Edge Gateways: Installed on each floor to normalise BMS, HVAC, and sensor data for energy monitoring and predictive maintenance.

- Landlord Portal: Built a common portal displaying building-wide analytics, alerts, and tenant reports, accessible via the cloud.

Results: Enabled proactive fault detection and lifecycle planning across a fragmented estate.

Key Takeaway: Landlords and managing agents can retrofit smart overlays without disrupting tenants, creating a common data ecosystem for shared benefit.

Case Study 5: Repurposing a Vacant Floor for Hybrid Collaboration

Location: Watford, UK

Building Age: Early 2000s

Challenge: A floor left vacant post-pandemic due to downsizing, with a desire to test hybrid meeting formats and reduce building-wide energy use.

Transformation:

- Modular AV Pods: Installed preconfigured smart pods with integrated lighting, ventilation, and AV control.

- Zoned HVAC Control: Introduced occupancy-driven HVAC logic, heating only active zones.

- Digital Twins: Used Revit and IoT data overlays to create a digital twin for scenario testing and future-fit planning.

Results: Reduced energy usage by 40% on the floor and reactivated the space as a hybrid collaboration and events area.

Key Takeaway: Smart zoning and modular systems enable rapid, cost-effective repurposing of dormant space into high-value hybrid areas.

Case Study 6: From Silicon Roots to Smart Rebirth — Retro-Smarting a San Jose Tech Office

Location: San Jose, California, USA

Building Age: 1992

Challenge: A former tech company HQ built during the dot-com boom had fallen behind on energy performance, lacked modern collaboration spaces, and had outdated wiring and AV infrastructure. The goal was to reposition the space for modern hybrid teams and attract new tenants.

Transformation:

- Layered Smart Infrastructure: Introduced an IWMS (Integrated Workplace Management System) with a focus on real-time data from HVAC, occupancy, and lighting systems, using BACnet and MQTT-compatible gateways for integration.

- Hybrid-Ready Huddle Zones: Reconfigured open-plan zones into bookable huddle areas equipped with Cisco Webex Room Kits, presence-sensing AV, and intelligent light/temp presets triggered by booking data.

- LEED Re-Certification: Added solar-powered window shading, IoT air quality sensors, and machine learning-driven HVAC scheduling to support a push for LEED Gold certification.

Results: Reduced carbon footprint by 36%, increased occupancy by 40%, and significantly improved tenant satisfaction.

Key Takeaway: American retrofits often prioritise energy performance, certifications (e.g., LEED), and tenant attraction, demonstrating the commercial value of modernising with smart layers and integrated control.

Case Study 7: Smart Transformation in a Subtropical Setting — Singapore's Government Office Refresh

Location: Singapore

Building Age: Late 1990s

Challenge: A public sector building in Singapore's CBD faced issues with energy inefficiency, overcrowding, and an inflexible layout that couldn't accommodate new hybrid work policies. Strict government procurement and data policies added complexity.

Transformation:

- IoT and BMS Fusion: Leveraged Singapore's Smart Nation initiatives by integrating a centralised BMS with new IoT layers for air quality, thermal comfort, and real-time footfall. Open-source standards like KNX and BACnet ensured scalability across departments.

- Geo-Aware Wayfinding and Booking: Deployed an advanced wayfinding and desk booking app linked to employees' government ID badges, streamlining workplace access and space utilisation.

- AI-Driven Maintenance: Embedded analytics triggered automated service tickets for facilities issues (e.g., temperature fluctuations or low lighting), reducing downtime and improving service levels.

Results: Reduced energy use by 25%, improved hybrid work readiness, and established a reference model now adopted by other ministries.

Key Takeaway: In smart city hubs like Singapore, government-led initiatives can accelerate smart building transformations, with a strong focus on interoperability, user experience, and AI-powered operational efficiency.

Conclusion

Each of these adaptive transformations reflects a common truth: legacy buildings are not obsolete—they're latent smart spaces waiting for the right spark. By tailoring technology choices to the physical, operational, and budgetary realities of each site, organisations can unlock real value from their estates. The key lies in starting with what's possible, layering in technology strategically, and viewing constraints as creative catalysts rather than barriers.

Whether it's a heritage landmark or a standard 90s office block, the smart retrofit revolution is here—and it's more attainable than ever.

In the US, certification, commercial returns, and hybrid collaboration tools dominate.

In Asia, especially in smart cities like Singapore, the focus includes government policy alignment, intelligent automation, and deep integration into civic digital infrastructure.

Each example proves that regardless of the geography, smart retrofits are achievable with the right mix of modular technology, creative planning, and a deep understanding of local drivers.

Part 3: End-to-End Execution

Chapter 17: Phased Delivery & Pilots

Rolling out smart workplace systems across an entire building or estate is rarely a single-step journey. Instead, most successful implementations begin with targeted pilots, evolve through incremental layers of technology, and culminate in a fully orchestrated environment over time. This phased approach reduces risk, allows for continuous learning, and provides stakeholders with visible value early in the process. Aligning the delivery roadmap with the 360 Smarter Stack methodology ensures each stage is logically grounded, from foundational infrastructure through to seamless user experience.

Designing a Roadmap Aligned to the 360 Smarter Stack

Delivering a smart workplace isn't simply a matter of installing sensors and dashboards—it's a strategic transformation that must align with a company's purpose, people, operations, and physical environment. The 360 Smarter Stack offers a powerful framework for sequencing initiatives in a way that ensures clarity, coherence, and long-term sustainability. Its eight interconnected layers create a roadmap that balances early value delivery with a scalable and future-ready foundation.

Why the Stack Matters

By aligning a smart delivery roadmap to the 360 Smarter Stack, organisations avoid common pitfalls—such as tech-first deployments that fail to gain traction, siloed system rollouts, or infrastructure that can't scale. Instead, they progress logically from strategic intent to tangible impact, allowing flexibility to test and adapt while still ensuring overall cohesion.

Layer 1: Vision & Value

This is the critical starting point—defining why you're pursuing smart transformation. Is the goal to improve energy efficiency, enhance employee wellbeing, support hybrid work, reduce operational cost, or align with sustainability and ESG goals? Once the "why" is clear, it drives every subsequent decision.

Key activities in this phase include:

- Establishing programme goals and business drivers.

- Setting success metrics and KPIs.

- Engaging senior stakeholders through a strategic steering group.

- Developing a robust business case, potentially tied to OPEX/CAPEX rebalancing.

- Articulating benefits in human, environmental, and financial terms.

This layer sets the tone: no amount of technology will succeed without strategic clarity and alignment.

Layer 2: Operations & Culture

Smart offices will change how teams work—and how they're supported. This layer ensures that existing processes, roles, and service models are aligned with future needs. It also addresses potential friction caused by new tools or ways of working.

Key actions include:

- Readiness assessments of IT, FM, workplace, and support teams.

- Service model updates (e.g. integrating digital concierge or helpdesk tools).

- Role definition: who owns AV support? Who triages booking issues?

- Upskilling facilities and IT staff on new tools and platforms.

- Early-stage change management communications and town halls.

- Embedding workplace strategy into BAU operations.

A successful smart workplace doesn't just install tech—it supports it with the right people, processes, and culture from day one.

Layer 3: User Experience

Rather than forcing users to adapt to technology, smart offices should adapt to user needs. This layer is focused on mapping common user journeys—like arriving, booking a space, hosting visitors, or finding a colleague—and identifying friction points or opportunities for enhancement.

Typical tasks include:

- Conducting user journey mapping sessions with employees, FM, IT, and business reps.

- Creating service blueprints showing the front-end experience and back-end enablers.

- Identifying touchpoints that can be digitised or streamlined.

- Testing interaction design in pilot spaces (e.g. signage, kiosks, mobile apps).

- Gathering feedback and iterating before scale.

If the user experience is clunky or confusing, adoption will stall—no matter how clever the underlying tech is.

Layer 4: Applications & Services

Now that user needs are mapped, this layer focuses on selecting and deploying the right platforms to meet them. Key focus areas include booking systems, AV control, mobile workplace apps, energy dashboards, and visitor management.

Actions might include:

- Selecting best-fit software platforms with open APIs.

- Avoiding 'point solutions' in favour of ecosystems or suites that integrate well.

- Piloting core services like space booking and signage.

- Integrating platforms to deliver compound workflows (e.g. booking a room turns on HVAC and lighting).

- Building a long-term application architecture roadmap that supports flexibility and interoperability.

This is the layer where the digital front-end takes shape—and where integration opportunities are surfaced.

Layer 5: Data & Context

Data gives the smart office its intelligence. This layer ensures the organisation defines what data is important, how it's collected, and how it's used to create context-aware services.

Essential actions include:

- Tagging building assets, zones, and data streams using a consistent schema.

- Defining semantic layers to support analytics and automation (e.g. what is a "meeting room" across platforms?).

- Setting up pipelines to collect data from sensors, booking tools, energy meters, and user feedback systems.

- Creating dashboards with real-time and historical views.

Ensuring data privacy, security, and lifecycle management policies are in place.

This is where insights start to drive smarter operations—and where the groundwork for automation and AI is laid.

Layer 6: Integration Layer

Without integration, smart systems become isolated. This layer connects the various applications and systems, allowing them to interact through APIs, middleware, and logic engines.

Key components include:

- Selecting an integration platform (e.g. Microsoft Power Platform, Node-RED, or commercial middleware).

- Developing reusable workflows (e.g. room booking triggering AV setup).

- Establishing a central orchestration hub or digital twin layer.

- Ensuring time-synchronised and interoperable data exchange across systems.

The integration layer turns static systems into responsive, coordinated services that enhance both user experience and operational efficiency.

Layer 7: Systems & Devices

Here, we get hands-on with the physical technologies: sensors, cameras, displays, access readers, HVAC systems, lighting, and AV. The goal is to ensure the technology is robust, configured to spec, and connected to your integration and data layers.

Activities include:

- Selecting smart-ready devices with open protocols (e.g. MQTT, BACnet, DALI).

- Commissioning and testing in pilot zones.

- Creating device configuration standards and maintenance checklists.

- Ensuring all hardware is patched, secure, and remotely monitorable.

This is often the most visible part of a smart office—but it must be supported by every previous layer.

Layer 8: Physical Environment

Finally, all smart capabilities must be supported by physical readiness. The environment needs sufficient power, structured cabling, connectivity, mounting infrastructure, and flexibility to evolve over time.

Key focus areas:

- Ensuring rack space, cooling, and PoE switching capacity.

- Upgrading M&E as needed (e.g. DDC systems, lighting relays).

- Designing flexible meeting spaces and touchdown areas.

- Planning for device density and future scalability.

- Adapting older spaces for modern infrastructure needs.

Physical space is the final puzzle piece—where everything else comes together.

Bringing It All Together

A 360 Smarter Stack-aligned roadmap is not a strict checklist—it's a strategic compass. It allows organisations to:

- Prioritise based on readiness across layers.

- Pilot without breaking the bigger picture.

- Scale with confidence, avoiding technical debt or rework.

- Engage the right stakeholders at each phase.

- Move at a pace that balances innovation with operational stability.

Selecting a Pilot Space or 'Smart Zone'

Strategically testing the smart office vision in a controlled, measurable environment.

Once your roadmap is aligned to the 360 Smarter Stack, the next critical move is implementation—and that starts with a pilot.

Rather than going all-in across the entire estate, a smart pilot space—also called a "smart zone"—lets you test ideas, gather feedback, refine configurations, and validate value. It's the proving ground where technology meets people, and where bold ideas meet real-world friction. A successful pilot sets the tone and pace for scalable deployment across your portfolio.

Why Start Small? The Value of a Pilot Approach

A well-designed pilot delivers several strategic advantages:

- Risk reduction: Test new tech, processes, and user interfaces in a low-impact zone before committing.

- Cost control: Focus your investment on a contained space with high visibility and clear outcomes.

- Stakeholder confidence: Demonstrate tangible benefits to execs, FM, IT, and business leads through real usage data.

- Iterative improvement: Learn from feedback, adjust workflows, and refine configurations before scaling.

- Change management: Involve early adopters who can become champions during wider rollout.

Smart pilots aren't just "trials"—they're precision tools for de-risked innovation and scalable transformation.

Criteria for Selecting a Smart Pilot Zone

Not all spaces are equal when it comes to piloting smart capabilities. The ideal smart zone is a microcosm of broader organisational activity, but with some strategic advantages:

1. Representative Use Case

Choose a space that reflects key business activities—such as collaboration, hybrid meetings, visitor hosting, or focused work. Examples include:

- A floor with a mix of meeting rooms, desks, and breakouts.

- A shared collaboration hub used by multiple teams.

- A customer-facing area like a reception or executive lounge.

You want to validate your technology against real, diverse, and meaningful user journeys.

2. High Visibility & Engagement

Ideally, the space should be used by multiple departments or leadership stakeholders. This ensures:

- Greater exposure to new features.

- Broader feedback collection.

- Higher stakeholder interest and support.

Bonus: if senior leaders use the space, it helps fast-track buy-in for future phases.

3. Technically Feasible

Make sure the space:

- Has the necessary power, cabling, and network capacity.

- Allows for retrofit of devices (mounting, HVAC access, ceiling grid compatibility).

- Isn't undergoing other disruptive works (e.g. refurbishments).

- Can be supported with IT and FM resources for testing and incident response.

The best ideas will fall flat if the infrastructure can't support the vision.

4. Manageable Scope

Keep the space and system count contained but meaningful:

- 4–6 rooms and a set of desks, not an entire building.

- A single floorplate or "neighbourhood."

- Limited system integrations at first (e.g. room booking + lighting + occupancy).

This allows you to focus on experience, quality, and feedback—rather than firefighting.

5. Feedback-Ready

Make sure the space includes:

- A group of engaged users willing to participate in trials.

- Feedback collection tools, like QR codes, surveys, or embedded app prompts.

- Support channels for issue escalation (e.g. a dedicated IT/FM response team or digital concierge).

User insight is the most valuable output from the pilot—build mechanisms to capture it.

What Should Be Included in the Pilot?

Your smart zone should cover at least 3–4 layers of the 360 Smarter Stack to be meaningful—but not so many that it becomes unmanageable. Here's a common pilot configuration:

Smarter Stack Layer	Example Pilot Implementation
Vision & Value	Pre-agreed success metrics (e.g. utilisation uplift)
User Experience	Room booking, check-in displays, mobile app
Applications & Services	Desk/room booking, visitor system, digital signage
Data & Context	Occupancy sensors, utilisation dashboards
Integration Layer	Booking triggers lighting, AV presets
Systems & Devices	Touch panels, sensors, cameras, lighting relays
Physical Environment	Ceiling-mounted sensors, digital signage screens

Pilot Phases and Best Practices

A smart pilot isn't "set it and forget it"—it needs structure, evaluation, and iteration. Here's a recommended lifecycle:

1. Discovery & Definition

- Finalise success criteria (quantitative and qualitative).

- Confirm scope, budget, and technical dependencies.

- Identify key users and support staff.

2. Installation & Commissioning

- Fit out the space with smart systems and connectivity.

- Test integrations (booking to lighting, AV sync, etc.).

- Train the support team and early users.

3. Active Use & Observation

- Monitor adoption, incidents, and engagement.

- Use analytics platforms to track space usage.

- Collect feedback via prompts, interviews, or workshops.

4. Review & Optimisation

- Identify friction points, bugs, and unmet needs.

- Adjust signage, workflows, or policies as needed.

- Document lessons learned and update future rollout plan.

5. Showcase & Decision Point

- Create a summary report with outcomes and recommendations.

- Host walkthroughs for stakeholders and leadership.

- Decide on go/no-go for scaled rollout and plan the next wave.

Pilots as Strategic Accelerators

When well-executed, smart pilots build:

- Credibility with stakeholders.

- Confidence in the technology.

- Clarity on what works—and what doesn't.

- Culture change by involving early adopters.

- A road-tested blueprint for future implementation.

Whether you're retrofitting an existing office, launching a new hub, or preparing for a multi-site transformation, the smart zone model offers a practical, low-risk, high-insight way to start.

Chapter 18: Project Delivery Best Practices (Aligned with Project Phases)

Delivering smart office projects is as much about **people and process** as it is about technology. Mapping the **360 Smarter Stack** to traditional project phases provides a powerful delivery framework, but the true success comes from embedding **governance, roles, and stakeholder accountability** at each stage.

Whether rolling out a single floor or a multi-site global programme, this chapter focuses on the practical delivery methods that ensure alignment, transparency, and momentum.

360 Smarter Stack vs Project Phases

1. Initiation

Stack Layers:

- *Vision & Value* → Define the *why*: ROI, ESG goals, productivity outcomes.
- *Operations & Culture* → Identify stakeholders, governance, support readiness.

Governance:

- Executive Steering Group signs off the business case.
- Governance framework agreed (reporting cadence, decision rights, escalation).

Key Roles:

- Sponsor → champions business case, secures funding.
- Programme Manager → defines scope, ensures alignment.
- IT & Facilities Leads → validate feasibility.

Stakeholders:

- C-Suite (vision/ROI).
- Workplace, HR, IT, FM (operational readiness).

Purpose: Align strategy, set measurable goals, secure leadership buy-in.

2. High-Level Design

Stack Layers:

- *User Experience* → Map broad journeys, personas, touchpoints.
- *Applications & Services* → Define scope (booking, signage, dashboards, security).
- *Data & Context* → Outline high-level data strategy and integration needs.

Governance:

- Design Authority convened to approve principles.
- High-level requirements signed off, with scope control agreed.

Key Roles:

- UX Lead → owns journey mapping.
- Enterprise Architect → defines service boundaries.
- Data Architect → proposes integration/data flows.

Stakeholders:

- End-user reps (employees, visitors).
- Business units impacted (HR, IT, FM).

Purpose: Set design vision, clarify functional scope, avoid over-engineering.

3. Detailed Design

Stack Layers:

- *User Experience* → Detailed flows, accessibility, inclusivity.
- *Applications & Services* → Translate into technical specs.

- *Data & Context* → Define semantic model, tagging, standards.
- *Integration Layer* → Architect APIs, middleware, workflows.

Governance:

- Technical Design Authority reviews specs.
- Change Control Board manages scope changes.

Key Roles:

- Solution Architects → produce detailed designs.
- System Owners → validate functional requirements.
- Integration Specialist → maps interfaces.

Stakeholders:

- IT security, compliance, procurement.
- FM operations (maintenance, lifecycle planning).

Purpose: Ensure everything is build-ready, compliant, and fully specified.

4. Build / Test

Stack Layers:

- *Integration Layer* → Develop/test interfaces.
- *Systems & Devices* → Deploy HVAC, lighting, AV, access control, IoT.
- *Physical Environment* → Deliver cabling, power, racks, PoE.

Governance:

- Delivery Board oversees milestones and quality gates.
- Testing governance: UAT scripts, sign-offs, defect tracking.

Key Roles:

- Project Managers → oversee builds.

- Vendors/Installers → deliver systems.
- Test Manager → coordinates test cycles.

Stakeholders:

- Facilities → verify installs.
- IT Ops → validate integration and security.

Purpose: Deliver and validate the technical foundation before go-live.

5. Implement (Go-Live + Handover)

Stack Layers:

- *User Experience* → Validate journeys, signage, apps.
- *Applications & Services* → Deploy and drive adoption.
- *Operations & Culture* → Train IT/FM/support; align SLAs.
- *Vision & Value* → Track KPIs vs business case.

Governance:

- Cutover Board manages go-live risks and approval.
- Handover governance: knowledge transfer, BAU acceptance checklists.

Key Roles:

- Change Manager → manages comms and adoption.
- Training Lead → delivers role-based training.
- Service Manager → ensures support readiness.

Stakeholders:

- End-users (adoption focus).
- Facilities/IT Support (day-to-day operations).

Purpose: Ensure smooth cutover, embed systems, enable early adoption.

6. Close

Stack Layers:

- *Vision & Value* → Confirm benefits and lessons learned.
- *Operations & Culture* → Full BAU transition and updated governance.
- *Data & Context* → Validate reporting pipelines for future insights.

Governance:

- Formal Project Closure Report.
- Benefits Realisation Review.

Key Roles:

- Sponsor → signs off closure.
- PMO → archives documentation.
- Data Analyst → verifies metrics/KPIs.

Stakeholders:

- All governance boards (for lessons learned).
- Operational teams (final handover confirmation).

Purpose: Prove ROI, embed lessons, and prepare for future phases.

With governance and roles mapped, this becomes not just a methodology but a **repeatable delivery framework**. It tells teams **what to do, who does it, and how accountability flows** at every stage.

Interdependency Mapping Across Phases and Layers

Smart office delivery is not linear — each decision made in one layer or phase has a ripple effect across others. To manage this complexity, interdependencies must be mapped not just by **stack layer**, but also across **project phases**.

Example: Readiness Matrix by Phase and Layer

Phase	Layer	Inputs Needed	Outputs Provided	Risk If Skipped
Initiation	Vision & Value	Exec sponsorship, ROI targets, ESG goals	Business case, strategic KPIs	Project may lose direction or funding
High-Level Design	User Experience	Personas, workplace journeys	Draft UX maps, requirements	Features risk being irrelevant or poorly adopted
Detailed Design	Data & Context	System data sources, compliance policies	Data model, interoperability standards	Reporting & analytics won't function as intended
Build/Test	Integration Layer	APIs, middleware, system specs	Validated workflows, cross-system communication	Systems may not talk to each other, causing failure at go-live
Build/Test	Systems & Devices	Cabling, racks, procurement	Deployed IoT, AV, HVAC devices	Hardware delays or failures block adoption
Implement	Operations & Culture	Training plans, comms, champions	Users onboarded, BAU ready	Poor adoption, increased support calls
Close	Vision & Value	Post-occupancy data, lessons learned	Benefits realisation, exec reporting	Value not captured, undermining future investment

A "Readiness Matrix" is a practical tool for this purpose. It highlights what inputs each layer requires, what outputs it provides, and the risks of proceeding without it.

Use this as a dependency risk analysis tool at phase gates to sequence Go/No-Go decisions and ensure readiness.

Tools to Manage Interdependencies

Delivering a smart office requires more than technical coordination — it demands **governance tools** that give stakeholders transparency, align delivery to strategy, and avoid costly surprises.

Here are three of the most effective:

1. Change Logs: Tracking Evolution Transparently

Change is inevitable in smart programmes. A structured change log helps track, evaluate, and communicate adjustments across multiple layers.

- Use a shared, version-controlled register (e.g., SharePoint, Confluence, Jira).
- Record: impacted layer/phase, description, reason, status, owner, and impact analysis.
- Review changes in governance forums to maintain alignment.

2. Stakeholder Maps: Roles, Influence, and Engagement

With IT, FM, HR, and end-users all playing roles, mapping stakeholders ensures no group is overlooked.

- Categorise by function (e.g., IT, FM, HR, Exec).
- Align stakeholders to relevant stack layers.
- Map influence: "Consult," "Involve," "Inform," etc.
- Update regularly at major milestones or when scope shifts.

3. Integrated Gantt / Stack View: Phased + Layered Delivery

Traditional Gantt charts lack the nuance needed for layered delivery. An integrated Gantt/Stack view overlays **time** with **stack layers**, making interdependencies explicit.

Benefits:

- Shows sequencing: e.g., you can't validate UX until the Integration Layer is tested.

- Supports phased rollout (pilots vs. global).

- Highlights risks and blockers across phases.

Tools:

- Smartsheet, MS Project → timeline + dependencies.

- Miro, Figma → collaborative early-stage mapping.

- Power BI, Tableau → overlay reporting and dashboards.

Example: Layered Project Timeline

- **Phase 1 – Initiation:** Vision & Value, Ops & Culture (Days 0–20)

- **Phase 2 – High-Level Design:** UX, Apps, Data (Days 10–35)

- **Phase 3 – Detailed Design:** UX, Apps, Data, Integration (Days 20–55)

- **Phase 4 – Build/Test:** Integration, Devices, Physical (Days 30–70)

- **Phase 5 – Implement:** UX validation, Ops & Culture training, Apps deployment (Days 50–90)

- **Phase 6 – Close:** Benefits realisation, BAU transition, reporting (Days 90+)

Note: Timings are illustrative — real-world projects may span months or years.

360 Smarter Stack Aligned Gantt Chart (Illustrative)

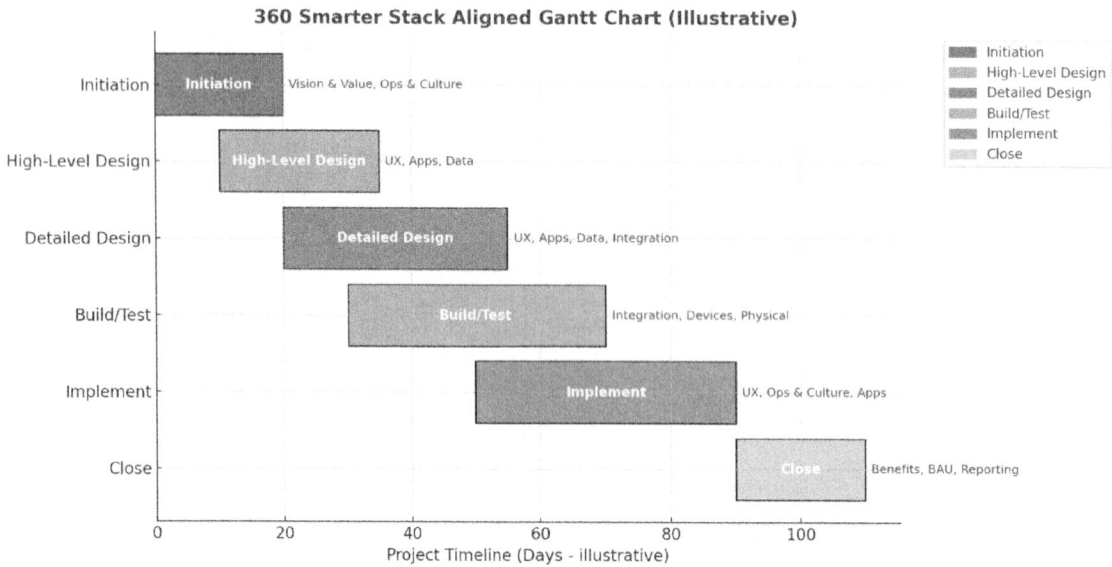

Diagram: Gannt Chart Showing Project Delivery Mapped To 360 Smarter Stack

Together, these tools form a backbone for structured project delivery in the complex, fast-evolving landscape of smart office design. They not only keep teams aligned and informed but create an audit trail of key decisions, pivots, and dependencies— supporting long-term governance and learning for future phases. Please note that the duration shown is purely indicative for illustrative purposes only, in reality projects can be significant in duration.

Chapter 19: Post-Go-Live and Beyond

The successful activation of a smart workplace solution does not mark the end of the journey — it marks the beginning of real-world performance, continual tuning, and value realisation. The post-go-live phase is where governance transitions from **project mode** (milestones, dependencies, risk registers) into **operational mode** (service levels, KPIs, and continuous improvement).

A well-structured post-go-live approach ensures that the smart office is **embedded sustainably**, user adoption is maximised, and the organisation is set up for iterative evolution rather than one-off delivery.

Handover and Training

From Project Governance to Operational Governance

At the end of the **Implement phase** (Chapter 18), governance must formally pivot. The project board hands authority to BAU structures, with defined **RACI matrices** for ownership across Facilities, IT, Security, Workplace Experience, and HR. This avoids the common "handover cliff" where knowledge is lost or accountability is unclear.

Key governance actions during handover:

- **Formal sign-off** of all deliverables and documentation packages.
- **Updated RACI matrix** showing who owns what in BAU.
- **Handover workshop(s)** with all stakeholder groups.
- **Change log closure** — freezing project changes and passing open requests into a "Day-Two Backlog."

Structured Handover Documentation

(aligns to governance traceability and dependency mapping from Ch. 18)

- As-built documentation (architectures, racks, devices)
- Configuration logs (settings, policies, permissions)
- Integration records (API endpoints, workflows)

- Maintenance manuals and SLAs

- Change logs and final approvals

These must live in a version-controlled repository (e.g., ITSM system, SharePoint with permissions).

Training by Audience and Role

Training ensures that governance isn't just written on paper — it's understood and actionable. Each stakeholder group must be **enabled to own their part of the stack**:

- **Facilities** → Smart HVAC, lighting, access, dashboards

- **IT** → Integration, cybersecurity, patching, SLAs

- **Reception** → Visitor/wayfinding, room visibility

- **End-users** → Apps, booking tools, signage interactions

- **Executives** → Dashboards, KPIs, ROI metrics

Delivery should mix **live sessions, microlearning, LMS modules, and QR-code quick guides** at physical touchpoints.

Champions and Train-the-Trainer Models

To embed culture and resilience:

- **Super Users** in each team trained at advanced level.

- **Workplace Champions** drive peer adoption and feedback.

- **Support escalation pathways** mapped to governance (service desk → IT/FM → vendor).

Snagging, Tuning, and Optimisation

Snagging

Where Chapter 18's Gantt/Stack ensured sequencing, **snagging checks that dependencies actually work in practice**. Issues logged here often reflect gaps between theory and operation (e.g., an integration tested in isolation but failing in live use).

- Track via CAFM/CMMS or Jira/Trello.

- Prioritise by severity, business impact, and user experience.

- Governed via weekly "hyper-care" reviews in first 30 days.

Tuning

Iterative adjustments ensure the system delivers **comfort, efficiency, and adoption**:

- HVAC setpoints informed by occupancy analytics.

- Lighting schedules adjusted for hybrid patterns.

- Access policies streamlined for evolving work models.

- Wi-Fi/AV rebalanced after live feedback.

Tuning governance typically shifts into BAU, but initial adjustments should be **monitored by the same cross-functional steering group** that owned delivery.

Optimisation

This is where governance evolves from defect-fixing to **continuous improvement**. Data-driven decisions inform:

- AI/automation enhancements.

- UX refinements from user feedback.

- Energy efficiency improvements (shutting down unused floors).

- API reliability or latency reviews.

Optimisation should be managed via a **Quarterly Review Cycle** with stakeholder representation across all layers of the 360 Smarter Stack.

Planning for Day-Two Enhancements

Why Day-Two Planning Is Governance-Critical

Chapter 18 established dependency and sequencing discipline. Day-two governance ensures that new features, integrations, or policies are introduced without destabilising existing systems.

Common Day-Two Enhancements by Layer

Stack Layer	Typical Day-Two Enhancements	Governance Requirement
Vision & Value	Re-align strategy with post-occupancy learnings	Exec steering review
Operations & Culture	Onboarding packs, refined policies	HR/FM ownership
User Experience	UX/UI refinements, accessibility upgrades	User councils/workshops
Applications & Services	Visitor mgmt, signage expansion	IT/Procurement sign-off
Data & Context	Enhanced analytics, external data sources	Data governance board
Integration Layer	New APIs/workflows	IT architecture review
Systems & Devices	More sensors, hardware upgrades	IT/FM sign-off
Physical Environment	Layout tweaks, AV/lighting upgrades	FM/Projects oversight

Governance and Ownership for Day-Two

- **Steering Group Oversight** (quarterly, cross-functional).
- **Day-Two Backlog** maintained in ITSM/PM tool.
- **Smart Tech Champion** role for advocacy and coordination.
- **BAU integration** — embedding optimisation tasks into IT/FM daily operations.

Communication and Transparency

Change governance isn't complete without **clear comms**:

- Announce new features via signage, intranet, or email.
- Share "what's new" dashboards to show progress.
- Celebrate adoption wins and efficiency savings.

Part 4: The Future of Smart Offices

Chapter 20: Scaling and Evolving the Stack

As organisations mature in their smart building journeys, the focus shifts from individual deployments to enterprise-wide transformation. The 360 Smarter Stack provides a strong foundation, but its long-term success hinges on the ability to scale, adapt, and evolve with new technologies. Emerging innovations like digital twins, artificial intelligence (AI), and ambient intelligence are reshaping how buildings learn, respond, and serve occupants. Equally critical are considerations for cybersecurity and the challenge of orchestrating consistent experiences across a global estate. This chapter explores how to future-proof your smart office investments and scale them sustainably across portfolios.

Digital Twins, AI/ML, and Ambient Intelligence

Digital twins are virtual representations of physical environments, systems, or assets. They enable real-time visualisation, simulation, and optimisation. By integrating live data from sensors, building management systems, and user feedback, digital twins act as living dashboards—allowing facility managers and IT teams to test scenarios, predict failures, and optimise space and energy usage.

When combined with AI and Machine Learning (ML), digital twins become proactive. Machine learning algorithms can detect usage patterns, predict maintenance needs, and automate responses. For example, by learning occupancy trends, an AI-powered twin might adjust heating and lighting dynamically, improve air quality during busy hours, or pre-emptively alert engineers to faulty equipment.

Ambient intelligence builds on this by embedding contextual awareness into the environment. It uses AI, sensor fusion, and advanced analytics to create spaces that respond autonomously. Lights dim as natural daylight increases. Room booking systems unlock spaces as users approach. Access control adapts based on role, time of day, or location. Together, these technologies deliver truly responsive environments.

Benefits include:

- Reduced energy consumption through predictive adjustments.

- Enhanced user comfort and productivity.

- Faster response to faults or anomalies.

- Increased asset longevity via predictive maintenance.

To succeed, organisations must invest in foundational data infrastructure and ensure consistent data tagging, semantics, and governance across assets.

Cybersecurity in Connected Environments

As buildings become increasingly smart and interconnected, they also become attractive targets for cyber threats. From intelligent lighting and access control to HVAC systems, sensors, and digital signage, each connected device represents a potential entry point into your corporate network. In this new landscape, cybersecurity must be treated as a fundamental architectural layer, not an afterthought.

The Expanding Attack Surface

In traditional IT environments, network perimeters were well-defined. But in smart buildings, operational technology (OT)—which includes building management systems (BMS), IoT devices, and embedded controllers—converges with IT networks. This convergence introduces new vulnerabilities:

- Legacy devices with outdated firmware or default credentials.

- Unencrypted communications between devices and cloud systems.

- Lack of segmentation between systems, allowing lateral movement by attackers.

- Shadow IT—unauthorised devices or systems added without proper oversight.

A cyberattack targeting a smart office might aim to disable HVAC systems during a heatwave, unlock secure doors, intercept data from occupancy sensors, or launch ransomware via compromised display systems. Beyond operational disruption, these threats can compromise personal data, damage reputations, and result in regulatory penalties.

A Multi-Layered Defence Strategy

Securing connected environments requires a layered, proactive approach:

1. Network Segmentation:

- Separate IT and OT networks using VLANs and firewalls. Isolate critical building systems from general-purpose traffic and apply strict access controls.

2. Zero Trust Architecture:

- Adopt a "never trust, always verify" mindset. Authenticate and authorise every device, user, and application based on role, context, and behaviour.

3. Secure Device Onboarding & Management:

- Ensure all devices are provisioned through secure channels, use encrypted communication (TLS/SSL), and support remote firmware updates. Disable default credentials and unnecessary services.

4. Monitoring & Threat Detection:

- Deploy real-time monitoring and intrusion detection systems (IDS) to observe traffic anomalies, scan for vulnerabilities, and flag suspicious activities across both IT and OT layers.

5. Governance & Compliance:

- Align your smart building cybersecurity strategy with global standards and frameworks such as:

    1. IEC 62443 – for industrial control systems security.

    2. NIST Cybersecurity Framework.

    3. ISO/IEC 27001 – for information security management.

6. Vendor Risk Management:

- Evaluate the cybersecurity posture of third-party vendors providing smart building components. Include security clauses in contracts and require regular updates, patches, and audits.

7. Regular Penetration Testing & Audits:

- Simulate attacks on your smart systems to identify vulnerabilities before malicious actors do. Include both digital (software/network) and physical (on-site access, device tampering) vectors.

Building a Culture of Security

Technology alone isn't enough. A secure smart workplace depends on the human element. It's vital to:

- Educate FM, IT, and procurement teams on cyber hygiene and risk.

- Foster collaboration between cybersecurity, IT infrastructure, and workplace experience teams.

- Establish incident response plans tailored to building systems.

- Ensure change management processes include cybersecurity reviews.

Looking Ahead

As smart buildings evolve, so too must cybersecurity strategies. The rise of edge computing, AI-driven attack tools, and autonomous systems will introduce new complexities. Future-ready organisations will embed security by design, not only protecting assets but also enabling innovation with confidence.

A secure smart office is not only resilient—it's trusted, agile, and prepared for the challenges of a hyper-connected future.

Scaling the Smarter Stack Across Global Portfolios

Once a smart building strategy has proven successful at one site or region, the natural next step is to scale those capabilities across a broader real estate portfolio. Whether you're managing a national network of campuses or a global estate of corporate offices, scaling the 360 Smarter Stack requires more than duplication—it demands standardisation, localisation, and governance at enterprise scale.

From One to Many: The Strategic Challenge

Scaling smart technologies across sites introduces new challenges:

- Varying building ages, layouts, and infrastructure readiness.

- Inconsistent local regulations and compliance standards.

- Diverse vendors, platforms, and legacy systems already in place.

- Cultural and operational differences between regions or teams.

Without a coherent and scalable approach, organisations risk ending up with a patchwork of disconnected smart buildings—each a silo, each requiring separate management and support.

Key Enablers of Scalable Smart Estates

To avoid fragmentation and ensure long-term scalability, organisations must adopt a repeatable but flexible blueprint. Here's how:

1. Define a Global Smarter Stack Reference Architecture

Create a standardised architecture model—based on the 8-layer 360 Smarter Stack—that can be adapted to local contexts. This acts as a "gold standard" guiding every deployment across:

- Layer 1: Vision & Value – Define global objectives (e.g. ESG, digital experience, cost reduction) while allowing local KPIs.

- Layer 2: Operations & Culture – Develop playbooks for FM and IT teams with room for regional process nuances.

- Layer 3 to 8 – Ensure standardisation of platforms, integration approaches, sensor types, security protocols, and interface design.

- Think of it as a "kit-of-parts"—a modular, adaptable framework applied consistently.

2. Establish Governance and Ownership Structures

Successful scaling depends on clear accountability. Consider:

- A central Smart Building Programme Office to lead architecture, vendor strategy, and funding models.

- Regional Smart Champions or Delivery Leads to localise rollouts and liaise with site teams.

- A global governance board overseeing milestones, approvals, security, and performance.

Define who owns each layer of the stack at each stage—design, build, operate—and assign named roles to avoid gaps.

3. Embrace Platform Thinking, Not Just Projects

Rather than treating each site as a standalone project, implement shared platforms for:

- Device management (IoT hubs, provisioning platforms).

- Centralised analytics and dashboards.

- Identity and access control integration.

- Maintenance and support processes.

This allows for remote support, global data insights, and reduced cost through economies of scale.

4. Create a Repeatable Deployment Playbook

Develop a scalable, phase-based delivery playbook that includes:

- Readiness assessment templates (infrastructure, data, user needs).

- Pilot and Proof of Concept strategies for high-impact areas.

- Commissioning checklists for all systems and integrations.

- Training and handover frameworks to ensure consistency.

Where possible, use digital twins or 3D site models to streamline planning and simulation before boots hit the ground.

5. Harmonise Data and Reporting Standards

To truly scale smart capabilities, data must be consistent, accessible, and contextualised. That means:

- Standardising data tagging and semantic models across all sites.

- Creating a common data lake or integration layer.

- Designing dashboards that allow for cross-site benchmarking (e.g. energy usage per m², occupancy trends, indoor air quality scores).

This approach empowers global teams to compare, learn, and optimise—while avoiding false comparisons.

6. Prioritise Cybersecurity at Scale

Cybersecurity becomes more complex at scale. Adopt global frameworks for:

- Device provisioning and certificate management.

- Role-based access controls aligned across regions.

- Ongoing monitoring and incident response with central oversight.

- Audit trails and compliance alignment with local laws (e.g. GDPR in Europe, CCPA in California).

Security breaches in one region shouldn't compromise the rest of your estate.

7. Measure Success with a Living Maturity Model

Develop a global Smart Building Maturity Model that measures progress across:

- Deployment of stack layers

- Operational effectiveness

- User satisfaction

- Environmental impact

- Business outcomes

Use this model to benchmark buildings, highlight success stories, and identify underperforming sites.

A Vision for the Global Smart Portfolio

Scaling the 360 Smarter Stack globally isn't about creating identical buildings—it's about enabling intelligent consistency with local relevance. The outcome is a smart portfolio where:

- Data flows seamlessly across sites.

- Experiences are intuitive and unified, regardless of region.

- Improvements compound over time as learnings from one site inform the next.

- Cost and complexity reduce, while innovation accelerates.

By applying rigour, flexibility, and foresight, global organisations can move from smart buildings to smart portfolios—and ultimately, to intelligent ecosystems where every square metre adds value to people and performance.

Conclusion: Unlocking Smart Potential, One Layer at a Time

The journey through the 360 Smarter Stack has provided a layered, practical framework for envisioning, planning, deploying, and evolving smart building initiatives. Whether you're retrofitting a legacy office or launching a flagship intelligent workspace, this methodology ensures that digital transformation is structured, scalable, and meaningful at every stage.

A Recap of the 8-Layer Stack

Each layer plays a crucial role in shaping the success of smart environments:

- Vision & Value – Establishes the 'why', aligning initiatives with business goals, user needs, and long-term outcomes.

- Operations & Culture – Embeds change readiness, governance, and collaboration across teams to ensure sustainable adoption.

- User Experience – Focuses on intuitive, human-centred design—whether it's digital signage, room booking, or space navigation.

- Applications & Services – Covers the digital tools, platforms, and interfaces that empower occupants and operators.

- Data & Context – Turns raw data into meaningful insights through real-time analytics, dashboards, and predictive intelligence.

- Integration Layer – Ensures interoperability using APIs, middleware, and standards like MQTT, BACnet, or KNX to stitch systems together.

- Systems & Devices – Includes the physical hardware—sensors, AV, access control, HVAC—that deliver the smart capabilities.

- Physical Environment – Considers the built infrastructure: cabling, racks, PoE, and M&E readiness to support reliable system performance.

Bringing It All Together

From roadmap planning and pilot zones to Gantt charts, stakeholder tools, and governance dashboards, each chapter has aimed to provide both strategic insight and tactical guidance. The 360 Smarter Stack isn't just a blueprint—it's a working model that adapts to context, grows with your portfolio, and evolves with the technology landscape.

Whether you're focused on sustainability, user productivity, operational efficiency, or brand experience, the Stack gives you the clarity to act, the tools to build, and the foresight to scale.

The Road Ahead

Smart building technology is rapidly advancing—AI, machine learning, digital twins, and ambient intelligence are no longer buzzwords; they're becoming foundational to real estate and workplace strategy. Meanwhile, cybersecurity, governance, and data privacy continue to demand attention as systems grow more connected.

With the 360 Smarter Stack as your guide, you're equipped to navigate these complexities with confidence. You now have a structure for aligning people, technology, and space—layer by layer, step by step.

Appendices

Appendix A: Smart Office Stack Checklist

Use this checklist as a structured guide to assess your progress and coverage across the 360 Smarter Stack.

1. Vision & Value

✓ Clear business objectives and success metrics defined

✓ Executive sponsorship and stakeholder alignment secured

✓ Smart building goals tied to workplace strategy or ESG targets

✓ Value proposition understood across departments (IT, FM, HR, etc.)

✓ Long-term roadmap established (3–5 years)

2. Operations & Culture

✓ Internal champions and cross-functional governance group established

✓ Staff training and change management plan in place

✓ Operating model defined for managing smart tech post-go-live

✓ Maintenance roles (FM/IT) clarified for shared responsibilities

✓ Feedback loops embedded for user input and continuous improvement

3. User Experience

✓ Defined and mapped user journeys (visitors, employees, facilities, etc.)

✓ Touchpoints audited (signage, booking, navigation, etc.)

✓ Unified front-end (mobile app, kiosk, portal) strategy confirmed

✓ Accessibility and inclusivity principles applied

✓ Pilot user group testing conducted and documented

4. Applications & Services

✓ Workplace experience app/platform selected and integrated

✓ Room and desk booking tools operational

✓ Wayfinding, visitor management, and feedback tools deployed

✓ Applications integrated with back-end systems (HR, BMS, etc.)

✓ License usage and performance monitored

5. Data & Context

✓ Real-time and historical data sources identified

✓ Semantic model or tagging strategy developed

✓ Occupancy, energy, and environmental metrics monitored

✓ Data warehousing or data lake established if required

✓ Contextual insights used in FM, energy, or space planning decisions

6. Integration Layer

✓ APIs mapped and documented across all platforms

✓ Middleware or integration platform (iPaaS) in place

✓ Standards in use (MQTT, BACnet, KNX, REST, etc.)

✓ Systems synced with enterprise apps (e.g., SAP, Office 365)

✓ Alerts and events configured across subsystems

7. Systems & Devices

✓ IoT sensors deployed and calibrated (occupancy, air quality, etc.)

✓ AV, lighting, HVAC, and access control systems connected

✓ Asset register created with device IDs and software versions

✓ Firmware upgrade paths and remote access secured

✓ Smart systems commissioned and acceptance tested

8. Physical Environment

✓ Cabling and rack layout designed for modularity and expansion

✓ PoE budget calculated, and switch density planned

✓ Comms rooms secured, and environmental conditions monitored

✓ Conduits, containment, and floor boxes coordinated with M&E

✓ Electrical and mechanical resilience (UPS, failover) checked

Final Readiness Check

✓ Pilot zone successfully tested and reviewed

✓ Snagging list closed out and post-handover support in place

✓ Governance and escalation paths defined for BAU

✓ Day-two enhancements scheduled (roadmap in place)

✓ KPIs and insights reporting in operation

Appendix B: Stakeholder Workshop Tools

Engaging stakeholders early and effectively is key to the success of a smart building programme. The templates below are designed to help you structure and facilitate productive workshops across departments, project phases, and layers of the 360 Smarter Stack.

1. Vision & Value Workshop

Purpose: Align strategic goals, define outcomes, and agree success criteria.

Agenda	Tools
Introductions & Objectives	Icebreaker cards
Smart Building Vision Overview	Presentation slides
Business Drivers Discussion	SWOT Analysis (template)
Value Proposition Mapping	Value Canvas
Success Metrics Brainstorm	KPI Framework Board
Action Plan & Next Steps	Workshop Notes + RACI Chart

Key Questions:

- What does 'smart' mean to us?

- Which business problems are we solving?

- Who are the executive sponsors and beneficiaries?

2. User Experience Discovery Workshop

Purpose: Map user journeys, pain points, and digital touchpoints.

Agenda	Tools
Current State Journey Mapping	User Persona Templates

Common Frustrations & Quick Wins	Empathy Maps
Target Experience & Moments of Delight	Future-State Journey Canvas
Prioritisation & Feasibility Matrix	MoSCoW Board / Dot Voting
Wrap-up & Actions	Digital Collaboration Board (Miro)

Key Questions:

- What do users want to achieve each day?

- Where are their biggest friction points?

- What experience should our workplace deliver?

3. Integration Architecture Workshop

Purpose: Align IT, FM, AV, and BMS teams on platform architecture and interfaces.

Agenda	Tools
System Landscape Review	Integration Map (Editable)
Current State vs. Target State	Architecture Diagram Templates
API Inventory and Mapping	API Catalogue Template
Middleware & Data Flow Discussion	Swimlane Flow Diagrams
Integration Risks & Dependencies	RAID Log

Key Questions:

- Which systems need to talk to each other?

- What are the integration constraints?

- Are we aligned on protocols and data ownership?

4. Facilities & Infrastructure Workshop

Purpose: Plan for cabling, comms room layouts, M&E coordination, and device coverage.

Agenda	Tools
Physical Infrastructure Audit	Floorplan Mark-ups
Rack & Cabinet Planning	Rack Elevation Templates
PoE, Network & Sensor Budgeting	Port Count Calculator
Overlap with M&E (lighting, HVAC, etc)	Overlay Diagrams / BIM Viewer
Site Survey Walkthroughs	Photographic Log / Site Notes

Key Questions:

- Where are the key coverage zones for AV, Wi-Fi, sensors?

- Do we have enough capacity and resilience?

- What retrofitting constraints exist?

5. Governance & Change Workshop

Purpose: Define governance, roles, risks, and comms plan.

Agenda	Tools
Roles & Responsibilities	RACI Matrix
Change Impact Assessment	Change Canvas
Comms Planning	Stakeholder Map
Project Milestones & Gates	Milestone Wallchart
Risk & Readiness Discussion	RAID Register

Key Questions:

- Who signs off decisions at each layer?

- What needs to be communicated to whom, and when?

- How will change be monitored and approved?

Appendix C: RACI and Delivery Matrix by Layer

Clear ownership and accountability are essential for successfully implementing a smart office. Using a RACI (Responsible, Accountable, Consulted, Informed) model across the 8 layers of the 360 Smarter Stack helps align roles, avoid duplication, and identify any gaps in ownership. Below is a breakdown of how a RACI matrix can be applied to each layer, along with a delivery matrix showing typical stakeholders, timelines, and interdependencies.

Layer 1: Vision & Value

Role	Executive Sponsor	Workplace Strategy	IT Leadership	FM/CRE Lead	Smart Consultant
RACI	A / R	R / C	C	C	C

Delivery Notes:

- High-level alignment required early.
- Stakeholder buy-in is critical to shape the strategy and secure funding.

Deliverables: Vision statement, business case, success KPIs.

Layer 2: Operations & Culture

Role	FM/CRE Lead	HR/People Lead	Smart Consultant	IT Lead	Department Heads
RACI	A / R	R / C	C	C	C / I

Delivery Notes:

- Emphasis on change management, cultural readiness, and behavioural nudges.

Deliverables: Change strategy, communication plan, training programme.

Layer 3: User Experience

Role	UX Designer	FM/CRE Lead	IT Lead	End Users	Comms Lead
RACI	R	A	C	C	I

Delivery Notes:

- Strong engagement with end users for journey mapping and pain point identification.

Deliverables: User journey maps, personas, UI/UX requirements.

Layer 4: Applications & Services

Role	IT Lead	Product Manager	Workplace Tech PM	Vendors	Cybersecurity
RACI	A	R	R / C	C	C

Delivery Notes:

- Selecting interoperable platforms for booking, wayfinding, room control, etc.

Deliverables: App catalogue, platform integration strategy.

Layer 5: Data & Context

Role	Data Architect	Smart Consultant	IT Lead	Business Analyst	Cybersecurity
RACI	A / R	C	C	R / C	C

Delivery Notes:

- Designing contextual data layers, semantic models, and ensuring GDPR compliance.

Deliverables: Data dictionary, integration schema, ownership policies.

Layer 6: Integration Layer

Role	Solutions Architect	IT Lead	Vendors	API Developers	Smart Consultant
RACI	A / R	C	C	R	C

Delivery Notes:

- Middleware, APIs, and orchestration logic defined and implemented.

Deliverables: Integration plan, middleware stack, API documentation.

Layer 7: Systems & Devices

Role	M&E Lead	AV Specialist	IoT PM	BMS Vendor	Network Engineer
RACI	A / R	R	R / C	C	C

Delivery Notes:

- Covers sensors, smart panels, AV gear, HVAC upgrades, etc.

Deliverables: Device list, commissioning protocols, firmware/version tracking.

Layer 8: Physical Environment

Role	Architect	FM Lead	M&E Engineer	Construction Lead	Accessibility Advisor
RACI	A / R	C	R / C	C	C / I

Delivery Notes:

- Involves physical space modifications to support tech installations.

Deliverables: Layouts, accessibility plan, rack and containment plans.

Delivery Matrix Snapshot (Simplified)

Layer	Initiation	Design	Build	Commissioning	BAU Handover
1. Vision & Value	✓	✓			✓
2. Operations & Culture	✓	✓	✓	✓	✓
3. User Experience	✓	✓	✓	✓	✓
4. Applications & Services		✓	✓	✓	✓
5. Data & Context		✓	✓	✓	✓
6. Integration Layer		✓	✓	✓	✓
7. Systems & Devices		✓	✓	✓	✓
8. Physical Environment	✓	✓	✓	✓	✓

Appendix D: Business Case Calculation Example

Objective:

To provide a framework and worked example for calculating the ROI, payback period, and strategic value of implementing a smart office project across a defined office space (e.g., a 3,000 sq. m building).

Assumptions

Variable	Value
Office size	3,000 sq. m
Number of employees	300
Avg. desk occupancy	60%
Cost per workstation per year	£7,500
Energy cost per sq. m	£35
Smart office implementation cost	£1.2M upfront
Annual maintenance cost	£60,000
Expected annual benefits (Year 1–5)	Detailed below

Cost Breakdown (CAPEX & OPEX)

Item	Year 0 (Upfront)	Annual (OPEX)
Infrastructure upgrades (AV, HVAC, PoE)	£500,000	–
Applications, platforms, licenses	£250,000	£30,000
IoT sensors and integration	£150,000	£10,000
Design, consultancy, PM	£150,000	–
Training, comms, change management	£100,000	£20,000
Total	**£1,200,000**	**£60,000**

Projected Annual Benefits

Benefit Category	Estimate (£/Year)	Description
Space Optimisation	£375,000	Downsizing from 300 desks to 200 (100 x £7,500)
Energy Savings	£40,000	40% reduction in energy usage (3000 x £35 x 40%)
Maintenance Efficiencies	£25,000	Predictive maintenance reduces contractor visits
Staff Productivity Gains	£90,000	10 mins/day saved per user via smart tools (~300 FTEs)
Improved Meeting Efficiency	£20,000	Meeting room utilisation and automation gains
Total Annual Benefit	**£550,000**	**Cumulative Year-on-Year**

ROI & Payback Summary

- **Year 1 Net Benefit** = £550,000 – £60,000 OPEX = £490,000

- **Payback Period** = £1.2M / £490,000 ≈ 2.45 years

- **5-Year Net Benefit** = (5 x £490,000) – £1.2M = £1.25M

- **ROI Over 5 Years** = (£2.45M – £1.2M) / £1.2M = 104%

Strategic Value (Intangible Benefits)

Strategic Pillar	Value Add Description
Employee Experience	Enhanced satisfaction, retention, and attraction of talent.
Sustainability Targets	Measurable reduction in energy use and carbon footprint.

Brand & Innovation	Signals leadership in technology and sustainability.
Operational Resilience	Proactive monitoring and remote diagnostics reduce downtime.
Scalability	Architecture built to scale across other locations.

Sensitivity Analysis (Optional)

You may model different scenarios such as:

- **Conservative:** Only 50% of the estimated benefits realized
- **Aggressive:** Increased benefits from expanded use (e.g., AI-based occupancy predictions)
- Scenario with phased deployment across 5 sites

Summary Table

Metric	Value
Total Upfront Investment	£1.2M
Annual Ongoing Cost (OPEX)	£60,000
Annual Benefit	£550,000
Payback Period	~2.45 years
5-Year Net Benefit	£1.25M
ROI (5 years)	104%

Appendix E: Glossary of Smart Building Terms

A

AI (Artificial Intelligence):
The simulation of human intelligence by machines. In smart buildings, AI enables systems to learn patterns and automate functions like climate control, security, and space optimization.

Ambient Intelligence:
An environment where sensors, devices, and systems work together seamlessly to respond intelligently to occupants' needs and behaviours.

API (Application Programming Interface):
A set of protocols and tools for integrating software applications. APIs allow different smart building systems (e.g., HVAC and access control) to communicate and share data.

B

BACnet (Building Automation and Control Networks):
A data communication protocol widely used for integrating HVAC, lighting, and other building automation systems.

BMS (Building Management System):
A centralized platform used to monitor and control building systems such as HVAC, lighting, power, and security.

BYOD (Bring Your Own Device):
A policy allowing users to connect personal devices to the building's network, often used for personalized controls in smart environments.

C

CCTV (Closed-Circuit Television):
Used for video surveillance and security. In smart buildings, CCTV can be integrated with analytics for occupancy tracking or threat detection.

Commissioning:
The process of verifying and documenting that building systems are designed, installed, tested, and capable of being operated and maintained according to operational requirements.

Cybersecurity:
Protecting building systems, data, and networks from unauthorized access or cyber threats. A key concern for smart buildings due to increased connectivity.

D

Digital Twin:
A real-time digital representation of a physical building, system, or asset, used for simulations, monitoring, and optimization.

DALI (Digital Addressable Lighting Interface):
A standard protocol for intelligent lighting control that enables fixtures to be individually addressed and programmed.

E

Edge Computing:
Processing data closer to where it is generated (e.g., at a sensor or device) rather than sending it to the cloud, improving response time and reducing bandwidth.

Energy Dashboard:
A real-time interface displaying energy usage and performance data, often used in facility management.

F

Firmware:
Permanent software programmed into a device's memory. In smart devices, firmware updates can improve functionality or security.

G

Gantt Chart:
A visual project management tool that shows tasks against time. Used in smart office rollouts to track progress across stack layers.

Gateway:
A device that connects different systems or protocols, such as translating Zigbee signals into BACnet or IP for centralized control.

H

HVAC (Heating, Ventilation, and Air Conditioning):
A critical system for indoor climate control. Smart HVAC systems use sensors, AI, and analytics to optimize energy and comfort.

I

IoT (Internet of Things):
A network of interconnected devices embedded with sensors and software to exchange data over the internet or local networks.

Integration Platform:
Middleware or software that allows multiple building systems to interoperate, often through APIs or standard protocols.

L

LoRaWAN (Long Range Wide Area Network):
A low-power, long-range wireless protocol ideal for transmitting small data packets from IoT sensors in buildings or campuses.

Lighting Control System:
Automated systems for managing artificial lighting, often incorporating daylight sensors, motion detectors, and user preferences.

M

Middleware:
Software that connects different components of a tech stack (e.g., hardware and applications), facilitating data flow and communication.

Mesh Network:
A type of network topology where nodes (devices) communicate with each other directly, enhancing coverage and resilience.

O

Occupancy Sensor:
A device that detects motion or presence in a space, used to automate lighting, HVAC, or security systems.

P

PoE (Power over Ethernet):
Technology that allows network cables to carry electrical power, simplifying installation of devices like sensors, access control panels, and wireless access points.

R

RACI Matrix:
A project management tool outlining who is Responsible, Accountable, Consulted, and Informed for each activity or layer.

S

Smart Zone / Pilot Space:
A designated area within a building used to test and validate smart technologies before wider rollout.

Stakeholder Map:
A visual representation of individuals or groups involved in or affected by a smart building project.

Sensor Fusion:
The process of combining data from multiple sensor types to gain more accurate or insightful information (e.g., temperature + motion + CO_2).

T

Telemetry:
Data collected remotely from devices (e.g., temperature readings from sensors) and used for analytics, alerts, or automation.

Tuning:
Fine adjustments made to smart systems post-installation to ensure optimal performance.

U

Uptime:
The amount of time a system remains operational and available. High uptime is a key KPI in building resilience.

V

Vision & Value Layer:
The foundational layer in the 360 Smarter Stack, aligning smart office goals with business outcomes and leadership priorities.

W

Workplace Analytics:
Insights derived from data collected about how spaces are used, helping organizations make informed decisions on design and occupancy.

Z

Zigbee:
A wireless communication protocol often used in smart lighting and home/building automation devices.

Appendix F: Sources and Contributors

The following publications, frameworks, and tools informed the development of the 360 Smarter Stack methodology and its application across modern workplace projects:

- British Council for Offices (BCO) Guidelines – Best practice standards for office design and technology integration.

- ASHRAE Standards – Guidance for HVAC system design, sustainability, and building performance.

- WELL Building Standard v2 (IWBI) – Framework for health and well-being in the built environment.

- LEED v4 (U.S. Green Building Council) – Green building certification system.

- UK Government GDS Service Manual – Standards for digital project delivery, stakeholder engagement, and service design.

- CIBSE (Chartered Institution of Building Services Engineers) – Guidance on M&E systems, lighting, and smart infrastructure.

- Smart Buildings Alliance (SBA) – Resources on IoT integration and smart building governance.

- BSI PAS 212:2016 – Smart Cities – Semantics for data from smart appliances.
- Project Haystack – Standardised tagging and semantic modelling of data from building equipment.

- Digital Twin Consortium – Resources and use cases for digital twin applications.

- McKinsey & Company Smart Buildings Reports – Industry insights on digital transformation and smart building ROI.

- Deloitte Future of Work Framework – Influence on culture, flexibility, and technology alignment.

- Microsoft Azure IoT Hub & Digital Twins Documentation – Cloud integration reference models.

- Autodesk BIM 360 & Revit Documentation – BIM lifecycle support and integration references.

- Schneider Electric EcoStruxure – Smart building architecture inspiration.

- Siemens Desigo CC / Enlighted System – IoT sensor platform and AI integration examples.

- Honeywell Forge – Smart facilities management platform overview.

- MIT Smart Spaces Research Lab – Studies on behavioural analytics and responsive environments.
- IFMA Smart Building Benchmark Data – Operational metrics and industry benchmarks.

Afterword

The smart office is not a product — it's a journey

The vision of a smart office isn't something you can simply buy, install, and declare complete. It is not a product with a fixed endpoint — it's a dynamic, evolving journey that mirrors the way people, technology, and workplaces continue to change and adapt. As tools become more intelligent, user expectations more nuanced, and businesses more agile, so too must our approach to designing, deploying, and managing the smart environments in which we operate.

The 360 Smarter Stack provides a structure, not a finish line. It is a method to ask better questions, align across disciplines, and build with intent. But it is also an invitation to evolve — to revisit your vision, re-engage your stakeholders, and refine your technology as needs shift and opportunities emerge.

Smart buildings, and smart offices especially, thrive when they're treated as living systems. No deployment is ever "done." There will always be new integrations to explore, better experiences to deliver, data to uncover, and insights to act upon.

Keep Iterating. Keep Listening.

As you move forward in your smart office journey, remember that iteration is your greatest ally. Whether it's tuning an occupancy algorithm, revisiting your access control policies, or experimenting with a digital twin, every small step forward builds long-term resilience and value.

Success in this space doesn't come from perfection — it comes from momentum. From pilot to portfolio, from snagging lists to day-two enhancements, and from initial vision workshops to global scaling, the organizations that win are those who stay curious, stay collaborative, and stay committed to learning.

The smart office is a shared endeavour — across IT, facilities, design, leadership, and users. And its greatest power lies in the conversations it sparks, the silos it breaks, and the new experiences it makes possible.

Thank you for being part of this journey. Now go build the future — smarter, layer by layer.

Index

Printed in Dunstable, United Kingdom